FISH & FISHERIES DIGEST

4

POISONOUS & VENOMOUS FISHES AND SHELLFISHES

S. K. GUPTA
Retd. Associate Prof.
Department of Zoology,
DBS (PG) College, Dehradun (Uttarakhand), India.

Cover design: © Varun Gupta.
Colourful cover page is made lively out of watercolour brush strokes.
Visit (@shashivarungupta Instagram) to witness his watercolour and charcoal creations.

Copyright © 2022 S.K. Gupta

All rights reserved.

ISBN:

DEDICATION

TO
'The Inspiring Forces'

1.
AN INDIAN ICHTHYOLOGIST

Being an **'Ichthyological'** write-up, the present work is intended to be dedicated to the cherished memories of the great Indian Ichthyologist **Rai Bahadur Sunder Lal Hora** (2nd May 1896 – 8th December, 1955). Born at Hafizabad, West Punjab (Pakistan); post-graduated from Govt. College, Lahore; appointed assistant superintendent in-charge of Ichthyology and Herpetology at Zoological Survey of India in 1921 (ZSI); conferred on D.Sc. degree by Punjab University in 1928 for his work on 'torrential fishes'; appointed Director of Fisheries, Bengal in 1942; appointed as Director ZSI in 1947.

His prolific output began with a faunistic study on the 'Fishes of Seistan' in 1920. He wrote about 438 papers, including some monographs. Earlier he developed an interest in the role of fish in ancient Indian Culture. He proposed the famous **'Satpura Hypothesis'** regarding the distribution and dispersal of Malayan fish fauna in India etc.

Index Horana is the famous tribute by his one of the students Dr. KC Jayaram, where he synthesized what his beloved teacher achieved for Ichthyology. A number of honours were showered on Dr. Hora for his epoch-making work in the field of Ichthyology. It is my firm belief that no taxonomical or faunistic can go without referring to his classical publications, especially with reference to India.

2.
MY REVEREND FATHER
Late Shri O.P. Gupta, *Ex Principal*
Sadhu Ram Intermediate College, Dehradun, Uttarakhand, India

By virtue of being a devoted teacher and also an author of English and Biology Books, he often used to quote famous poet Milton:

"……*Books….* do contain a potency of life in them to be as active as that soul was whose progeny they are; nay they do preserve as in a vial the purest efficacy and extraction of that living intellect that bred them".

3.
MY WIFE 'VEENA'

"All other goods by fortune's hand are given: A wife is a peculiar gift of Heaven". – Pope

"Wives are young man's mistress; companions for middle age; and old man nurses". - Becon

Her sacrifices and memories are my strength to carry on with renderings in the field of education. To be busy with academic creations keep me afresh and all such products are submitted to her lingering remembrances.

- **SKG**

PREFACE

'Because books help people make their lives better in ways far beyond what's measured just by money' [*scribewriting.com*].

Such inspiring quotes have been zeal and zest-infusing to always do something for the cause of education at large. While discharging active services in a college, penning exercise by the author commenced way back in 1992 with the publication of **'An Introduction to Animal Behaviour (Ethology)'** [Bishen Singh Mahendra Pal Singh, Dehradun, Uttarakhand, India]. It remained continued until 2006 when another voluminous authorization of specific field, **'General and Applied Ichthyology (Fish & Fisheries)'** [S. Chand and Co., New Delhi, India] was brought-out for the benefit of the students pursuing Fish and Fisheries curricula. The said authorization received encouraging applaud from far and wide and thus was able to find prime place under suggested readings listed in the curricula of Fish & Fisheries courses, offered by various Indian Universities / Institutes.

Things did not end at this juncture and exploration of vast horizons in the field of publishing books led to bringing-out more authorizations in a **'SERIES'** format, on various topics of Fish and Fisheries. Not to be emphasized that the introduction of **'digital technology'** in book publishing, from kindergarten to Higher Education, has drastically changed the scenario of getting the 'Books' published for the voracious readers. Both the options, the online availability of **'eBooks'** or **'Paperback'** Editions, are now in tune with the working and study style of most present-day, digitally literate students and the like. In the changed internet-oriented, widely-adopted 'digital scenario', it is worth recalling that, '.........***books still have value in the 21st century, regardless of what form they take'***. [*https://study.com > Courses > Business Courses*]

A **'SERIES'**, entitled **'Fish & Fisheries Digest'** (in both formats) was, thus, launched in February, 2021 and its **Part - 1** and **Part – 2**, respectively entitled **'Fish & Fisheries - Monitoring and Management Systems'** and **'Integument and Colouration'** have already been at the hand of readers.

To maintain the 'subject/content connectivity', **PART – 3** was devoted to emphasize the significance of skin playing an important role in producing **'Light and Electricty'**, as evidenced by the following textual excerpts:

*"....... luminescence is emitted directly from the structures called as **'photophores'**, visible externally on the body. Made of **specialized gland cells of epidermis**, they are innervated and vascular structures, arranged in rows or in other patterns over the body............ .... This is an **epidermal tissue**, secreting a mucous layer which covers most of the head and body with red luminescence..."*.

*.... "..........showing bean-shaped suborbital light organ in **Photoblepheron** with a black shutter of **elastic skin** to be drawn up over the organ..........."* (From the pages of **Chapter – 1, PART-3** of the SERIES).

The interesting aspect of **'Biofluorescence'** also signified the **role of skin** and its components in causing fishes to fluoresce:

*"........ In majority of cases fluorescent **chromatophores** with fluorescent proteins (pigments) cause fluorescence. .... Like other cold-blooded vertebrates, in fishes too, chromatophores are the specialized cells in the **skin**, responsible for generating two types of colourations viz., structural and pigment-based. ....the pigment cells responsible for generating 'fluorescence' in fishes are the 'fluorescent chromatophores', occurring in **dermis** amongst other chromatophores. ...."* (From the pages of Chapter – 2, **PART-3** of the SERIES).

This 'episode', **PART – 4**, will prove to be more interesting in the sense that it will be unveiling another facet of **'skin'** by discussing a vast diversity of fishes/shellfishes, becoming enfamous by being **'Poisonous or Venomous'**.

It will be interesting to refresh from the textual matter of **PART - 4**, that noxious substances synthesized by specialized cells of the skin of some cartilaginous and bony fishes and the **venoms** produced by glands associated with spines on the fins or opercula not only provide protection to the fish from the predators but also empower them to subdue the enemy/predator.

Instances of fishes/shellfishes becoming **'poisonous'** to eat are not uncommon. Although, the 'poisonous' quality of fish/shellfish is mostly due to consumption of those fish/shellfishes which feed on poisonous algae/dinoglagellates/diatoms and get **'deadly toxins'** accumulated in their body, whether **skin** or **viscera**. The contents of this episode will discuss cases of *'**Fugu/puffer poisoning**'*, where it is suggested that viscera and **skin** must not be eaten under any circumstances.

Sometimes, the seafood (fish/shellfish) is mistaken as harmful by the body's immune system and this is reflected in the form of **'allergic reactions'**. The **Chapter-5** of this episode is, thus, devoted to **'Seafood Allergy'**, after giving proper attention to **'poisonous and venomous fishes and shellfishes'** in **Chapters 1** to **4**.

As usual, for the benefit of the 'competition-oriented' (NET/JRF) students (especially **India** and other countries of Southeast Asia), a 'Question Bank' is placed at the end of this part of the SERIES, too.

The following inspiring 'Quotes' will always pave way for 'penning' more for this 'SERIES':

"Put your heart, mind, and soul into even your smallest acts. This is the secret of success".
- Swami Vivekananda (the Great Indian Thinker)

"I can shake off everything as I write; my sorrows disappear, my courage is reborn".
- Anne Frank (German-Dutch diarist)

"Either write something worth reading or do something worth writing".
- Benjamin Franklin

" The most important thing to point out, is that despite lots of writing to the contrary, the book is not dead. …………Part of what I'm doing is advocating the importance of the book in history and as history".
-Michael F. Suarez, S.J, Director, Rare Book School, University of Verginia (UVA).

- S. K. Gupta

Table of Contents

INTRODUCTION 7
- I. TOXIN, POISON AND VENOM: 7
- II. FOOD TOXICOLOGY: 9
- III. FISH BIOTOXICOLOGY: 9
- IV. ICHTHYOTOXISM: 10
- V. ICHTHYOACANTHOTOXISM: 10
- VI. HISTORY: 10
- VII. MODE OF ACTION OF TOXINS/POISONS: 11
- VIII. SEAFOOD ALLERGY: 11

POISONOUS FISHES 14
- 1.1 HISTORICAL ASPECTS: 14
- 1.2 DISTRIBUTION OF POISONOUS FISHES: 14
- 1.3. THE DIVERSITY OF POISONOUS FISHES: 15
- 1.4. GENERAL LINE OF TREATMENT/PREVENTION OF FISH POISONING: 37

VENOMOUS FISHES 39
- 2.1 NUMBER OF VENOMOUS FISHES, A PHYLOGENETIC BASIS: 39
- 2.2 DISTRIBUTION/HABITAT PREFERENCES : 39
- 2.3 EPIDEMIOLOGY, IN GENERAL : 40
- 2.4 THE DIVERSITY OF VENOMOUS FISHES : 40

POISONOUS AND VENOMOUS SHELLFISHES 56

[MOLLUSCA] 56
- 3.1 BIVALVE SHELLFISH-POISONING: 56
- 3.2 GASTROPOD SHELLFISH-POISONING: 67
- 3.3 CEPHALOPOD SHELLFISH-POISONING: 71
- 3.4 VENOMOUS MOLLUSCS: 73

POISONOUS AND VENOMOUS SHELLFISHES 80

[XIPHOSURA AND CRUSTACEA] 80
- 4.1 XIPHISAURAN or HORSE-SHOE CRAB POISONING: 80
- 4.2 CRUSTACEAN SHELLFISH-POISONING: 81
- 4.3 VENOMOUS CRUSTACEAN: 87

SEAFOOD ALLERGY 89
- 5.1 ALLERGIC SEAFOOD ITEMS & SUBSTANCES: 89
- 5.2 SYMPTOMS: 90
- 5.3 DIAGNOSTICS: 91
- 5.4 TREATMENT: 91
- 5.5 PREVENTION: 91

QUESTIONS	92
APPENDIX – I	97
ABOUT THE AUTHOR	99

ACKNOWLEDGMENTS

While taking-up the task of expressing gratitude I always recollect, *"Your first step towards perfection is acknowledging your imperfections"* [Wahab H. Butt]. The heavenly-abode departed parents occupy the foremost rung of gratitude to be expressed. Their counseling phrases - **'work is worship'**, **'duty is beauty'** and **'hard work pays'**, do ever remind me to follow the discourse, how cumbersome it may be. Though saddened by the departure of my 'better-half' (Veena Gupta), amidst building-up of plans of writing hobby, her sacrifices remain a vigorous source of wisdom and encouragement. These departed, inspiring souls also remind, *"Books are letters in bottles, cast into the waves of time... Keep reading. Keep writing. Keep fighting. We're still here"* [Amal El-Mohtar, Max Gladstone].

Being software/computer literates, my worthy 'Twin' sons (Tarun & Varun) have taken-up the 'Twin Job' of designing and editing for **eBOOK** or **'Paperback'** formats. Besides accomplishing their job assignments, sharing my 'job of writing' is an emotional and far from being expressed acknowledgement. May such lineages that are caring and concerned about the dreams, their parents wish to enliven, be bestowed with all love and affection !

Additionally, the younger of the 'Twins', 'Varun', besides being an 'Architect' in Software discipline, is a good artist by hobby. The **'Cover Page'** illustrations of this episode, too, are designed by him by using water colour brush-strokes. Out of hobby, his more water colour creations may be seen by venturing into his website '*www.varunguptart.com*'(@shashivarungupta Instagram).

The **most-deserved credit** is hereby due, to be extended to the painstaking **'contributors/ authors'**, sharing their knowledge and accomplishments with the academia world-over, in the form of **'illustrations'** available on various '*websites*', utilized/modified herein for the benefit of the students and the like.

Old friends, students and the academia is another segment whose contribution to a teacher's life can never be under-estimated. They deserve the prime and utmost gratitude.

Last but not the least '*Amazon Kindle Direct Publishing*' has provided the platform to bring the dreams of publishing **eBook** and **'Paperback Edition'** - **'Fish & Fisheries Digest' SERIES** - a reality. Their efforts to provide an alternative to printed books and publish on-line are worth commendable.

- SK GUPTA

INTRODUCTION

"All things are poisonous and there is nothing without poison. Solely the dose determines that a thing is not a poison".
... *Paracelsus* (1493-1541), medieval physician and the 'Father of Toxicology'.

"Fish and guests in three days are stale".
... *John Lyly* (1554-1606), an English writer, poet, dramatist, playwright and politician, best known for his books Euphues, The Anatomy of Wit (1578).

"Fish is the only food that is considered spoiled once it smells like what it is".
... *P. J. O'Rourke* (1947 - 2022), American political satirist, journalist and author.

"Fish is meant to tempt as well as nourish, and everything that lives in water is seductive".
... *Jean Paul Aron* (1925 -1988), a French writer, philosopher and journalist.

"Until such fundamentals as the anatomical distribution of fish venoms have been determined, the pharmacological and chemical characterization of these compounds will continue to be unstudied".
... *Dr. Bruce Walter Halstead* (1920 - 2002), Treatise on venomous marine organisms.

The relationship between **fish** and man has continuously being witnessed since more than 4000 years (especially in Asia) when aquaculture practices (= the farming of fish, shellfish, crustaceans, seaweeds etc.) have been found to be the source of protein (containing all the 10 essential amino acids) requirements. Besides fish, other organisms from freshwater or sea have also been the source of protein since time immemorial. Historical evidences indicate that the harvesting, processing and consumption of **'seafood'** are ancient practices that date back to the beginning of the Paleolithic period, about 40,000 years ago. Other than fish, **'seafood'** most commonly includes **Crustaceans** (Shrimps, prawns, lobsters, crabs etc.), **Molluscans** (bivalves, gastropods and cephalopods) and **Echinodermates** (sea cucumbers, sea urchins and occasionally starfishes) and also marine algae.

'Shellfish' is a **culinary** and **fisheries** term for exoskeleton-bearing aquatic invertebrates, including various species **of Molluscs, Xiphisaurans, Crustaceans and Echinoderms**. Although most shellfish are harvested from marine resources, some are also found only in freshwater. Familiar **marine** molluscs used as food by human beings include many species of Bivalves (mussels, cockles, clams, oysters and scallops), **Gastropods** (snails etc.) and **Cephalopods** (squids, cuttlefish, octopuses). Among Arthropods, **Xiphisaurans** include the edible **horse-shoe** or **King crabs** whereas the edible crustaceans include **shrimps, lobsters, crabs and prawns**.

All are regarded as **'super food'**, owing to the presence of high level of **polyunsaturated fatty acids (PUFAs)**, especially *omega-3* fatty acids, as they feed on such marine microalgae that are the source of high level of **PUFAs**. They are regarded essential for normal development of neural, cerebral and visual functions.

However, besides the positive aspects, some **negative** and more **unpleasant relationships** are also important to be noticed. Fortunately, the incidences of the negative experiences are minor *e.g.*, many orders of fish/shellfish are known to be **'poisonous'** to human beings. Some cause **illness** or **death** when eaten; others have **stinging organs** that introduce **'venom'** into the prey/human beings. Together they cause hazards to divers, fishermen and bathers and to aquarists and diners who might come into contact with such fish/shellfish or their flesh.

The adjectives 'poisonous' and 'venomous' are often used synonymously but they have distinct meaning. The phenomenon of 'poisoning' involves four important components *viz*., the **cause, subject, effect** and the **consequence** of poisoning, respectively, correlated with the four elements *viz*., **poison, the poisoned organism**, the **injury** to the cells and **symptoms**.

[**NOTE: All the above referred aspects are elaborated in Chapters 1 to 4**].

I. TOXIN, POISON AND VENOM:

Toxin, Poison and **Venom** are synonymous terms frequently used for any substance that causes injuries to the health or destroys life when absorbed into animal systems.

(i) Toxin:

A 'Toxin' (L. *toxicus/toxikon* = poisonous, bow poison *i.e.,* poison used on arrow) is a **'poison'** produced by an organism or a **toxin** is of **natural** origin, chemically pure substance and clearly identifiable. **Natural toxins** are the compounds produced by living organisms, hence, biotoxins (Fig. A); naturally present in **fish/shellfish** and fishery products or accumulated by these animals after feeding on toxin-producing diatoms/dinoflagellates/algae or living in water contaminated by the toxins produced by such organisms. Besides the toxins/poisons produced by diatoms/dinoflagellates/algae, there are **mycotoxins** (from fungi), **phytotoxins** (from higher plants) and **zootoxins** (from animals). Chemically, they have diverse molecular structures (mainly proteins or mixture of proteins) and differ in biological function and degree of toxicity.

The deadly toxin, like **Tetraodotoxin** (TTX), is the product of some **'Bacteria'**. Similarly, **Paralytic shellfish poisoning** (PSP), which is commonly associated with consumption of shellfish, that have accumulated neurotoxins produced by marine dinoflagellates; has been found associated with **cyanobacteria** that produce PSP toxins.

Since we are concerned here with fish/shellfish toxicity/venomocity, **zootoxins** are to be understood. Based on the **mode of action** or production, **zootoxins** fall into three categories *viz.,*

- **Crinotoxins: Produced by specialized poison glands (chiefly in the skin) and released into the environment, by means of a pore (Fig. A).**

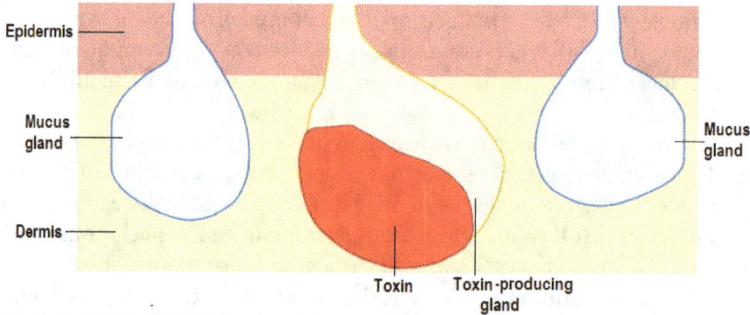

Fig. (A): Toxin, stored in the gland of the skin and delivered upon contact.

- **Oral poisons: They cause serious health problems only after ingestion by interacting with the bio-molecules within a living cell.**
- **Parenteral poisons or venoms: Produced by a venom apparatus composed of a specialized**
- **poison gland and injecting/inflicting device in the form of a proboscis, spine, teeth or sting (Fig. B).**

Toxins occurring naturally in diatoms/dinoflagellates/algae accumulate in fish/shellfish when they feed on them or on other fish that have fed on these diatoms/dinoflagellates/algae. Some of these natural toxins *viz.,* **tetraodotoxin, ciguatoxin, scombrotoxin, brevetoxins, okadaic acid, saxitoxins, ciguatoxin, domoic acid** etc, are the normal constituents of fish flesh. They result into serious health hazards *e.g.,* **'Fugu'** from the 'Puffers' (Tetraodontidae) has a lethal dose of **tetraodotoxin** (considered to be **about a thousand times deadlier than cyanide**) in its internal organs; **Ciguatera poisoning** is caused after eating sea bass, grouper and snappers from warm tropical waters; **Scombroid poisoning** (= Histamine or scombrotoxin) results after eating large, improperly frozen, oily tuna and mackerel-like fishes (Scombridae) etc.

On the other hand, some toxins are not of marine algal origin *e.g.,* **Gempylotoxin** (a strong purgative oil or wax ester) found in **Escolar** or **oilfish** (*Ruvettus* spp.) and **Tetramine** (= Tetramethylammonium salt) found in the salivary glands of a **Whelk Gastropod** (*Neptunea* spp.).

Many other hazardous poisonings *viz.,* Paralytic shellfish poisoning (PSP), Neurotoxic shellfish poisoning (NSP), Diarrhetic shellfish poisoning (DSP), Amnesic shellfish poisoning (ASP), Ciguatera

Fish Poisoning (CFP) and Azaspiracid Shellfish Poisoning (AZP) etc. are caused after consuming molluscans, shrimps, lobsters, crabs etc, having natural biotoxins accumulated therein.

The information regarding **biotoxins**, their pharmacological and chemical properties and the source of toxicity is not only vital for development of **antidotes** and rational assessment of the usefulness of the fish species as food source but it also opens a new arena for the exploration of biologically active chemical substances or biodynamic compounds of therapeutic value.

(ii) Poison:

The term **'Poison'** (Middle English *poison*, L. *potio/potionis* = drink, a draught, a poisonous draught, a potion = **a drinkinkable liquid that contains medicine** = poison or something that is supposed to have magic powers) is often used in a generic sense. In the language of biochemistry 'poison' is a natural or synthetic substance that causes damage to living tissues by way of injurious or fatal affect on the body, after being ingested or absorbed or injected.

To be more precise, a **poison** is a cocktail of various substances which are harmful for a body if they exceed a certain concentration *e.g.*, the poison of a bee consists of a variety of substances including toxins.

(iii) Venom:

The term **'Venom'** (L. *venemum* = magical charm, potent drug or deadly substance, poison) is a type of toxin /poison, secreted by an animal, equipped with a set of special apparatus having venom gland(s) for synthesizing venom and a proboscis/hollow or solid hypodermic teeth/ spine/sting for injecting/inflicting/delivering the deadly venom (Fig. B). Hence, venoms are often called **'parenteral poisons'** [Gk. *para* = beside + *enteron* = intestine] because of the fact that something is put inside the body through a device, but not by swallowing. Essentially, venom is a mixture of toxic/poisonous substances (proteins + other organic subastances) of different functional types *viz.*, cell-killing **necrotoxins** and **cytotoxins**, nervous tissue paralyzing **neurotoxins**, muscle-damaging **myotoxins** and blood coagulation obstructing **haemotoxins**. Further, venoms may contain both toxic and non-toxic components and a single organism, thus, can be both venomous and poisonous.

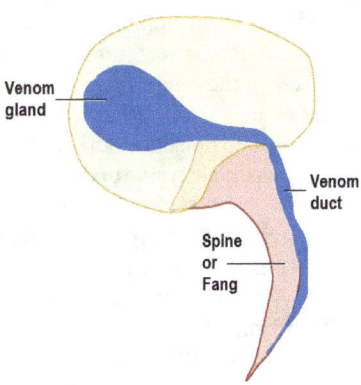

Fig. (B): Venom, produced in a gland connected to a specialized delivery apparatus, such as a fang/sting/spine.

II. FOOD TOXICOLOGY:

Various aspects of poisoning due to ingestion of fish/shellfish are dealt under a separate discipline, the **'Food Toxicology'**. It is the study of the nature, properties, effects and detection of **toxic substances** in food and their disease manifestation in human beings. Food Toxicology also includes the effect of environmental contaminants on the quality of flesh of fish/shellfish. **'Environmental Toxicology'** is another wider discipline, the primary focus area of which being studies on food toxicology, water quality, environmental contaminants (*e.g.*, mercury, Polychlorinated biphenyls or PCB's) and their effect on aquatic life (including fish /shellfish), human beings and wildlife.

III. FISH BIOTOXICOLOGY:

The science of **'Biotoxicology'** deals with knowledge about any organism by solving a number of complex biological and biochemical riddles yielding in several tangible results in terms of human health security and economic benefits. The studies in the field of **'Fish Biotoxicology'** have acquired new dimensions *viz.*, knowledge about the types of toxic fish/shellfish, nature and source of toxins and the conditions governing their toxicity etc.

As mentioned above, a number of **fish** and **shellfish** thrive on **biotoxin-producing** diatoms/ dinoflagellates/algae and other organisms. Toxins formed by algae in the ocean and freshwater are called **algal toxins**, being mostly generated during **'blooming'** of a particular algal species. Among scientific community, mass proliferation activity of toxic phytoplankton (particularly in the Oceans) is popular as **Harmful Algal Bloom(s)** [**HABs**] **[pl. refer to APPENDIX – I]**. There are about 10,000 beneficial marine phytoplankton distributed in the world's oceans and about 200 of them are adjudged producing harmful toxins often leading not only to human health hazards but also aquaculture fish-kills, thus, interfering the recreational activities in coastal or inland waters and causing great economic losses. Whales, porpoises, other aquatic and terrestrial animals also become victims when they accumulate toxins *via* contaminated water, plankton or fish.

Certain **marine biotoxins** *e.g.,* **ciguatera fish poison**, are the products of **marine plants** being transferred to herbivores and then to the carnivores and eventually to humans.

IV. ICHTHYOTOXISM:

On account of the involvement of potential 'toxins/poisons', derived from the food items which the fish consume, the related aspects are covered under the most conventional term **'poisonous fishes'** and this kind of poisoning is technically called **'Ichthyotoxism'**(= *ichthyism*). This poisoning/illness is caused to the consumers if the fish are not cooked/preserved properly. Therefore, **ichthyotoxism** is a kind of poisoning caused by a toxic substance derived from fish and the **'Ichthyotoxins'** are compounds, either toxic to fish or are toxins produced/secreted by fish or stored in their flesh. The fishes whose **tissues** are toxic, are called **cryptotoxic** (Gk. *cryptos* = hidden).

As per a classification given by **Dr. Bruce Walter Halstead**, the toxicity of fish through secretion and flesh consumption falls respectively, under two categories *viz.,* **ichthyocrinotoxic** (Gk. *ichthys* = fish; *krinein* = secretion) and **ichyosarcotoxic** (Gk. *ichthys* = fish; *sarcos* = flesh). More specifically, based on the kind of tissues where the toxins are identified *e.g.,* **blood** and **gonads**, the ichyosarcotoxic fishes are further classified as **icthyohaemotoxic** (Gk. *hemo* = blood) and **icthyootoxic** (Gk. *oo* = egg) fishes. The poisonous effect due to ingestion of **liver** is referred as **Ichthyohepatotoxic**.

Ichthyoallyeinotoxism (Gk. *ichthys* = fish; *aluein* = to be out of oneself, to hallucinate; *toxikon* = poison arrow), also known as hallucinogenic fish poisoning, is a kind of Ichthyosarcotoxism. It is a rare type **hallucinogenic intoxication** (a unique case of central nervous system ichthyotoxicity) due to the consumption of the head or other body parts of a fish.

All the cases of 'fish poisoning' referred herein are more aptly covered under **'Seafood Toxidromes'**.

V. ICHTHYOACANTHOTOXISM:

Scientifically, the term **Ichthyoacanthotoxin** (Gk. *acanth* = prickle or spine) is used to denote the tactics of delivering **parenteral poisons** or **venoms** through a specialized venom apparatus constituted by a venom gland and spines/stings/teeth. Such fishes with a definite venom apparatus are also called **phanerotoxic** (Gk. *phenoros* = evident).

VI. HISTORY:

The **Greeks** and **Romans** had a good knowledge of many naturally occurring poisons. Some written documents of about 450 BCE, describe the toxicity of venom released in snakebite and its possible treatment. In fact, death by poisoning was a common form of capital punishment *e.g.,* Socrates was forced to death sentence after drinking poisonous **hemlock** for supposedly corrupting the youth of Athens and failing to recognize official state deities.

The systematic study of poisons began as early as 16th Century, when the German-Swiss Physician and Alchemist **Theophrastus Bobmastus *von* Hohenheim** (*Paracelsus*) (1493 – 1541) first of all established the chemical nature of poisons. He introduced the concept of dose and studied the actions of poisons through experimentation and his valuable contributions established him to be known as the **founder** of modern **'Toxicology'**.

It was not until 19th century; however, that a Spanish Physician, **Mathieu Joseph Bonventure Orfila** (1787 – 1853), correlated **'chemistry of toxin'** with the biological effects it produces in the 'poisoned' individual. **Dr. Bruce Walter Halstead**, MD, (1920 - 2002) is regarded as the world's leading authority on poisonous fishes. He is better known as *'Newton Bruce Mellars'* due to his relationship with Sir Isaac Newton, an ancestor of his father. A bachelor's degree holder in Zoology, from the University of California (Berkeley) and a medical degree holder from Loma Linda University (USA), he opted for teaching as Assistant Director and Associate Professor of Preventative Medicine, School of Tropical and Preventative Medicine, Loma Linda University; as Assistant Surgeon, U.S.

Public Health Service; and as an Instructor in Tropical Medicine, U.S. Naval Medical School. His pioneer scientific investigations led him to the world of nature-based medicine *viz.,* fields of marine biotoxicology, toxic plants and animals of the world; tropical medicine, global pollution, chelation therapy (= administration of chelating agents to remove heavy metals from the body), hyperbaric oxygen therapy, medicinal plants, radiation sickness, AIDS, cancer, adaptogenic immune enhancement, nutrition and molecular biochemistry. On account of his expertise he earned the honour of serving as a consultant to more than 40 governmental and international agencies, including the World Health Organization (WHO), United Nations Educational, Scientific and Cultural Organization (UNESCO), U.S. Army, Navy and Air Force; the National Institutes of Health of various other governments etc. It was in 1995, when he conducted a major study for the U.S. Navy on the poisonous animals of the Mid-East, the results of which were compiled in the form of a book entitled *'Dangerous Aquatic and Land Animals of the Middle East'*. He is well known as the first one to establish the scientific field known as **'Marine Biotoxicology'**, due largely to his three voluminous treatises entitled, *'Poisonous and Venomous Marine Animals of the World'*.

VII. MODE OF ACTION OF TOXINS/POISONS:

Effect of **toxins** (= toxicity) can be **acute** (sudden), **chronic** (gradual) or both and in order to produce toxicity, sufficient quantity of toxic/poisonous chemical is to be absorbed into the body. Fundamentally, a chemical has to pass through cell membranes before entering into the blood or other tissues. The ability of a toxin/chemical to cross the lipid bilayer plasma membranes determines whether it will be absorbed and such ability depends on the chemical's lipid solubility. **Nonpolar toxins/chemicals** are **lipophilic** (lipid loving) and **polar toxins/chemicals** are **hydrophilic** (water loving). Lipid-soluble, nonpolar molecules pass readily through the membrane because they dissolve in the hydrophobic, nonpolar portion of the lipid bilayer. Although permeable to water (a polar molecule), the nonpolar lipid bilayer of cell membranes is impermeable to polar molecules, such as charged ions or those that contain many polar side chains. Polar molecules pass through lipid membranes *via* specific transport systems *viz.,* **diffusion**, **facilitated diffusion**, **active transport** and **pinocytosis**.

Although nonpolar chemicals cross the **skin** by diffusion through stratum corneum, no active transport exists in the dead cells of this layer. After a chemical crosses the transport barrier at the point of entry, it remains in the interstitial spaces between cells that are filled with water and loose connective tissue. The absorbed chemical gains access into the bloodstream directly *via* blood capillaries or indirectly *via* the lymphatic capillaries. The capillaries then drain into venules of the subcutaneous tissue. The porous nature of capillaries in most tissues or organs allows a chemical in the bloodstream to be distributed almost freely to most tissues, except for organs with a barrier. From interstitial spaces of the tissue, the molecules finally reach the cells either by diffusion or active transport. As a toxin/chemical is distributed to the tissues by the bloodstream, its concentration rapidly reaches a steady state with the blood concentration in the organs with profuse blood supply (liver, brain and kidney) while its concentration remains far below in the tissues with less profuse supply of blood (fat and bone). The rate of cutaneous absorption varies with the thickness of the stratum corneum at different sites of the body and the absorption is faster when the **skin** is moist rather than dry.

The toxins, eliciting various types of **toxic responses**, are classified by the nature of the response, the site of toxic action, the time it takes for the response to develop and the chance of resolution of the response. The nature of the toxic response can be **morphological** (structural) or **functional** or both. In most cases, the toxin/chemical produces morphological changes in an organ, which in turn affect the function of the organ. In a few cases, the toxin/chemical produces functional changes without changing the structure of the organ.

The **site of toxic response** can be at the site of first contact or portal of entry of the chemical (*i.e.,* **local**) or produced in a tissue other than at the point of contact or portal entry (*i.e.,* **systemic**).

VIII. SEAFOOD ALLERGY:

Seafood allergy is prevalent in those areas where seafood is an integral part of the diet. Allergenic reactions to seafood products are caused by **ingestion** and **handling** and **inhalation**. Allergy is a kind of **hypersensitivity** or an overreaction of the immune system, often leading to **anaphylaxis** (Chapter – 1). These anaphylactic reactions are exhibited by a person to such substances present in the environment which are harmless to most people. **Seafood** includes both fish (*e.g.,* Salmon, Tuna, Mackerel, Sardines and Cod) and shellfish *e.g.,* Mussels, Clams, Octopus, Squids, Shrimps, Lobsters and Crabs), responsible for severe *IgE*-mediated reactions, with **fish being one of the top eight allergenic foods**.

The filamentous muscle protein **tropomyosin** was first identified as a **shrimp allergen** and it is now also registered as a **fish allergen**. Skin reactions such as **urticaria** (hives) or **eczematous contact dermatitis** are the chief sysmptoms.

[NOTE: All the above referred aspects of 'SEAFOOD ALLERGY' are elaborated in Chapter - 5].

Chapter - 1
POISONOUS FISHES

The 'Puffer' or the '*Fugu*' (*Tetraodon, Takifugu sp.*), a 'watercolour illustration'.

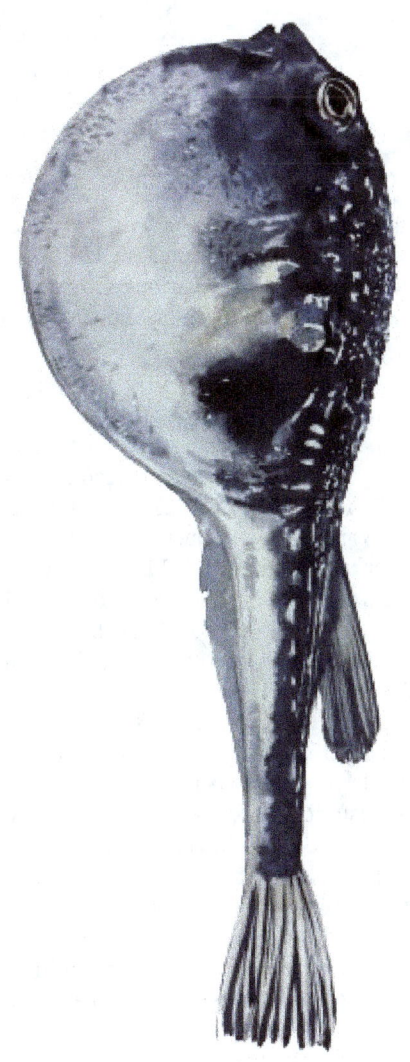

CHAPTER 1

POISONOUS FISHES

HIGHLIGHTS
HISTORICAL ASPECTS
DISTRIBUTION OF POISONOUS FISHES
THE DIVERSITY OF 'POISONOUS' FISHES
Ichthyocrinotoxic fishes

Ichtyosarcotoxic fishes
Ichthyohepatotoxic, Ichthyootoxic, Ichthyohemotoxic, Scombroid Poisoning, Tetrodotoxin-Poisoning, Clupeotoxic Fish, Gempylotoxic Fish, Ichthyoallyeinotoxic fish, Ciguatera Fish Poisoning, Haff disease

As per **FishBase** data (June 2022), about 35,000 species and subspecies of fish are described. Some species are dangerous (*e.g.,* poisonous, stinging, shocking or biting) and have been of immense concern to the fishery managers, scientists and various Goverments in some parts of the world. A rough estimate enumerates at least 1200 species of poisonous/venomous fish.

1.1 HISTORICAL ASPECTS:

The following quote from **Deuteronomy** (14:9-10), explains as to how the poisonous (**ichyosarcotoxic**) fishes were eliminated from diet:

"These ye shall eat of all that are in the waters: all that have fins and scales shall ye eat: And whatever hath not fins and scales ye may not eat: it is unclean unto you".

The French Archeologist **Claude Gaillard** (1923) reported that hieroglyphics and figures of the deadly **Puffer** (*Tetraodon lineatus*) appearing frequently on ancient Egyptian tombs, was recognized as poisonous (**ichyosarcotoxic**) during the times of **Pharaohs** (the political and religious leader of Egyptians, holding the titles: 'Lord of the Two Lands' and 'High Priest of Every Temple'). Also, in **Diego de Landa Calderón's** (1524 – 1579, a Spanish Bishop of the Roman Catholic Archdiocese of Yucatán) famous *Relacion de Las Cosas de Yucatan* (1566), the lethal qualities of puffers are mentioned. That the flesh of the **moray eel** is dangerous to eat, has been reported by Greek Physician **Aelius Galen** (129 – 216 CE) in his *De Alimentis* (1833). **Alexander the Great** (323 – 356 BCE), warned his soldiers not to eat fish during conquests because he believed that some species produced skin disorders. **Peter Martyr**, the first historian of the West Indies, made the earliest reference to **ciguatera** in 1555. The earliest historical account of **Ciguatera Fish Poisoning** (CFP) dates back to 1601, reported from south coastal area of Mauritius Island (Indian Ocean).

By the beginning of 19th century, a vast amount of literature had appeared on the subject of fish poisoning. The public health significance of poisonous fishes was pointed out in a series of outbreaks reported from Midway, Johnston and the Line Islands in the Central Pacific Ocean (USA) between 1943 and 1946. Recent mass intoxications in the western Pacific have attracted the attention of the public towards poisonous marine organisms causing health hazards. A bulk of the outbreaks have also been reported from Japan, the Philippine Islands and more recently in **Vietnam**. Most prevalent poisoning is **CFP** and it is estimated that about 10,000 - 50,000 cases are reported world wide, annually. Around 2014 the first epidemic of **CFP** in Germany highlights the increasing risk for other European countries.

As of March 2018, in two different fish poisoning cases, more than 50 people became the victim of fish poisoning in South Andaman region, after consuming Kutha Bhetki or Koduva Meen, the Asian Sea Bass or giant perch (*Lates calcarifer*).

1.2 DISTRIBUTION OF POISONOUS FISHES:

Poisonous fishes are largely circumtropical but on occasions found in temperate waters, too. Most poisonous fishes are shore-dwellers rather than oceanic inhabitants and are mostly confined to coral reef belts. Large populations of poisonous fishes are known to occur in the central Indo-Pacific area and in the West Indies. They are more common near islands than along continental shores and may vary considerably in their distribution on an island. A species may be toxic in one part of the island but edible in another. Lagoon fishes are more likely to be poisonous than those living on seaward reefs.

1.3. THE DIVERSITY OF POISONOUS FISHES:

There are as many as **11 different categories** of poisonings by fishes containing a poison/toxin within their **skin**, musculature, viscera and slime (Table 1.1). The majority of poisonings are observed after the fish is ingested. Most of the toxins identified being accumulated in the fishes come *via* marine diatoms/dinoflagellates/algae on which they feed or *via* other fish in the food chain. On account of this, fishes are usually named on the basis of the effect of poisoning on the part/tissue of the body involved *viz.,* **Ichthyohaemotoxic, Ichthyohepatotoxic, Ichthyootoxic** *etc*.

Some toxins are normal constituents of certain fish and therefore, the poisonings are named on the basis of kind of group/fish involved *e.g.,* **Lamprey** or **hagfish poisoning, Elasmobranch poisoning, Moray eel poisoning, Gempylid diarrhea, Scombroid poisoning** (= Scombrotoxic or histamine poisoning), **Puffer poisoning** (= tetraodotoxin poisoning) *etc*.

Ciguatera Fish Poisoning (CFP) (Ciguatera Syndrome) is the most prevalent food-borne poisoning caused by eating fishes like barracudas, snapper, parrotfishes, groupers, triggerfishes *etc*. whose flesh becomes contaminated with toxins produced by some specific dinoflagellates *e.g., Gambierdiscus toxicus* (PLATE III). **Haff disease** is a very recent case of poisoning due to ingestion of certain Cyprinids.

1.3.1 Ichthyocrinotoxic fishes: Fish with poisonous epidermal gland secretions:

(a) The Fish/Group (Table 1.1; **PLATE I**):

Cyclostomata:
Petromyzonidae (Lampreys), Myxinidae (Hagfishes).

Osteichthyes:
Muraenidae (Moray Eels), Batrachoididae (Toadfishes), Serranidae (Sea Bass, Groupers), Ostraciontidae (Trunkfishes), Solidae (Soles).

(b) Epidemiology and Toxicity:

Crinotoxins are also called **mucus toxins** and as such this category includes those marine fish (except Petromyzontidae which is also found in freshwaters of Northern Hemisphere) which produce mucus mixed toxic chemicals in **epidermal glands** (Fig. A). The primary function of these secretions has been to exhibit antibiotic activity, protecting fish from a myriad of invading microorganisms in the marine environment. On occasions, these secretions prove to be lethal to other fish, marine animals and even human beings. The toxicity of these glandular secretions is either when they are released into the surrounding medium or when the **skin** is consumed.

In both **Petromyzons** and **Myxines**, the safe edibility is only possible after the slime secreting glands and the lining is completely wiped off.

The **Moray eels** (*Gymnothorax* sp.) are considered poisonous if eaten as its mucus is toxic. **Toadfishes** produce a skin toxin that keeps predators away. They are edible if prepared properly but must be skinned and cooked thoroughly. Some species have been found to possess a lesser amount of highly toxic **'tetrodotoxin'**.

Members of the **Family - Ostraciontidae** are generally considered non-toxic and regarded as a table delicacy in Japan. However, since 1990, there have been reports about **boxfish poisoning** to humans beings due to ingestion of cooked fish. Formerly **Ostracitoxin**, but later named **Pahutoxin** (*'pahu'* is the Hawaiian name for the boxfish) from a **Box fish** (*Ostracion lentiginosus* or *O. meleagris*) [PLATE-I, g] is the first **ichthyocrinotoxin** whose chemical structure has been fully characterized. It is a ***choline chloride ester*** of *3-acetoxypalmitic acid* that behaves similarly as the **steroidal saponins** found in echinoderms. When the **toxic mucus** is released, it quickly dissolves in the environment and negatively affects any fish in the surrounding area. It is **hemolytic** and also works

Ichthyocrinotoxic fishes: [a] Lamprey; [b] Hagfish; [c] Moray Eel; [d] Toadfish; [e] Sea Bass or Soapfish (*Pogonoperca punctata*); [f] Grouper; [g] Box Fish (*Ostracion lentiginosus*); [h] Smooth Trunkfish (*Lactophrys triqueter*); [i] Red Sea Moses Sole (*Pardachirus marmoratus*).

as a **neurotoxin**, affecting nerve endings and leading to paralysis. Thenafter, the victim suffocates. Recently, another ichthyocrinotoxin, ***deacetoxypahutoxin*** was isolated from the **smooth trunkfish** (*Lactophrys triqueter*) [PLATE-I, h]. **Peptidic** or **proteinaceous** ichthyocrinotoxins have been first isolated from the **sea bass** or **soapfish** (*Pogonoperca punctata*) [PLATE-I e] and named as **grammistins** (hence, the Subfamily - Grammistinae of Order Perciformes). It caused ciguatera-like effects in cats.

Table 1.1: A systematic list of Poisonous fish along with the possible types of poisonings caused (indicated by numbers against them from 1-11*).

CYCLOSTOMATA (Jawless fishes)	**Atheriniformes (Silversides)**
Myxinidae: Hagfishes **1***	Scomberesocidae: Sauries **5**
Petromyzonidae: Lampreys **1**	Cyprinodontidae: Killifishes **3**
CHONDRICHTHYES (Cartilaginous fishes)	**Scorpaeniformes (Mail-cheeked fishes)**
Elasmobranchii: Sharks and Rays **2**	Cottidae: Sculpins **3**
OSTEICHTHYES (Bony fish)	**Perciformes (Perch-like fishes)**
Acipenseriformes (Sturgeons)	Serranidae: Sea bass and Groupers **1, 2,9,10**
Acipenseridae: Sturgeons **3**	Pomatomidae: Bluefishes **5**
Lepisosteiformes (Gars)	Carangidae: Jacks, Scads and Pompanos **5, 10, 11**
Lepisosteidae: Gars **3**	Coryphaenidae: Dolphinfishes **5**
Anguilliformes (Eel-like fishes)	Lutjanidae: Snappers **10**
Anguillidae: Freshwater eels **4, 11**	Sparidae: Breams **2**
Muraenidae: Moray eels **1,4**	Mullidae: Goatfishes **9**
Congridae: Conger eels **4**	Kyphosidae: Rudderfishes **9**
Ophichthyidae: Snake eels **4**	Pomacentridae: Damselfishes **9**
Elopiformes (Tarpon-like fishes)	Mugilidae: Mullets **9**
Elopidae: Ladyfishes **7**	Sphyraenidae: Barracudas **10**
Albulidae: Bonefishes **7**	Labridae: Wrasses **10**
Clupeiformes (Herring family)	Scaridae: Parrotfishes **10**
Clupeidae: Herrings and Sardines **5,7**	Trichodontidae: Sandfishes **2**
Engraulidae: Anchovies **5,7**	Stichaeidae: Pricklebacks **3**
Salmoniformes (Salmon family)	Gempylidae: Snake mackerels or Escolars **8**
Salmonidae: Salmonids **3**	Scombridae: Mackerels, Tunas and Bonitos **2,5, 10**
Esocidae: Pikes **3**	Xiphiidae: Swordfishes **5**
Cypriniformes (Carps)	Acanthuridae: Surgeonfishes **9, 10**
Cyprinidae: Minnows **3**	Siganidae: Rabbitfishes **9**
Catostmidae : Bigmouth Buffalo **11**	**Pleuronectiformes (Flatfishes)**
Characiformes (Characins)	Soleidae: Soles **1**
Serrasalmidae: Pacu-manteiga, Tambaqui, Pirapitinga **11**	**Tetraodontiformes (Puffer-like fishes)**
Siluriformes (Catfishes)	Ostracionidae : Trunkfishes **1**
Ariidae: Catfishes **3**	Tetraodontidae : Puffers or Globefishes **1,6**
Ictaluridae: Catfishes **3**	Diodontidae : Porcupinefishes **1,6**
Siluridae: Common catfishes **3**	Canthigasteridae : Sharp-nosed fuffers **6**
Batrachoidiformes (Toadfishes)	Molidae : Molas **6**
Batrachoididae: Toadfishes **1**	
Gadiformes (Codfishes)	
Gadidae: Codfishes **3, 11**	

1. *Ichthyocrinotoxic 2. *Ichthyohepatotoxic 3. *Ichthyootoxic 4. *Ichthyohemotoxic 5. *Scombrotoxic 6. *Tetrodotoxic 7. *Clupeotoxic 8. *Gempylotoxic 9. *Ichthyoallyeinotoxic 10. *Ciguatoxic 11. *Haff disease

[**NOTE:** Poisonous fish types, serial numbered **2** to **11** are included under what are called '**Ichthyosarcotoxic**'; the poisonous fishes with serial **No. 1** are basically due to toxic secretions from the body *i.e.,* '**Ichthyocrinotoxic**']

A milky substance secreted at the bases of dorsal and anal fins of **Red Sea Moses sole** (*Pardachirus marmoratus*) [PLATE-I, i] contains a lipophilic peptide ***pardaxin***, exhibiting shark repellent properties, causing increased solute permeability at the gills. ***Pardaxin*** also acts as **neurotoxin**, that induces neurotransmitter release at subcytotoxic concentrations by both calcium-dependent and calcium-independent mechanisms and causes cell death by **necrosis** at higher concentrations.

(c) Symptoms:

Eating skin can lead to gastrointestinal complaints. Even mere contact causes skin irritation. If such glands are associated with rays, parenteral envenoming may result. The major symptoms have been severe muscle pains arising out of **rhabdomyolysis** (see **Haff disease**, ahead), usually accompanied with discharge of black urine and abnormal elevation of serum creatinine phosphokinase.

(d) Treatment:

Symptomatic (*i.e.,* a treatment directed toward the relief of distressing symptoms, as opposed to treatment focused on underlying causes and conditions).

1.3.2 Ichtyosarcotoxic fishes (Fish flesh poisoning):
(i) Ichthyohepatotoxic: Fish with Poisonous Livers

(a) Species/Group (Table 1.1):

- **Chondrichthyes, Elasmobranchii:**
  Squaliformes (Sharks):

Carcharhinidae (Requiem sharks), Dalatiidae (Sleeper sharks), Hexanchidae (Cow sharks), Isuridae (Mackerel sharks), Scyliorhinidae (Cat sharks), Sphyrnidae (Hammerhead sharks).

Rajiformes (Rays): Dasyatidae (Stingrays), Myliobatidae (Eagle rays), Rajidae (Skates), Torpedinidae (Electric rays).

- **Osteichthyes, Perciformes (Perch-like fishes):**

Scombridae (Japanese Mackerel), Serranidae (Sea Bass) [PLATE Ie], Sparidae (Porgy), Trichodontidae (Japanese sandfish).

(b) Epidemiology and Toxicity:

Sharks and **Perch-like** fishes possess heat-stable toxins in the liver. Most reports of poisoning due to consumption of **sharks** date back to the first half of the 20th century and even earlier. Cases of poisoning due to consumption of livers of perch-like fishes have mainly been reported from Japan. In **rays** the toxins are found in the musculature and viscera. The poisoning syndrome caused by the consumption of sharks and rays is sometimes also called **'elasmobranch poisoning'**. There are indications that **hypervitaminosis** of **Vitamin A** is responsible for such toxic effects.

A fish poisoning involving 188 hospitalizations occurred in November 1993, in Manakara (south-east coast of Madagascar), following ingestion of **Bull shark** (*Carcharinus leucas*) [Fig. 1.1]. Two liposoluble toxins were isolated from the shark liver and tentatively named **Carchatoxin -A** and **–B**.

(c) Symptoms:

Fig. 1.1: Bull shark (*Carcharinus leucas*)

Nausea, vomiting, fever and headache appear within 30 min to 12 hours after consumption. The face usually becomes flushed and oedematous and a macular rash develops. Large areas of skin may peel off. Desquamation may continue for about 30 days. Most of the acute symptoms disappear in about 3–4 days.

(d) Treatment/Prophylaxis:

Symptomatic.

(ii) Ichthyootoxic: Fish with Poisonous Roe

(a) Species/Group (Table 1.1):

- **Osteichthyes:**

Acipenseridae (Sturgeons), Lepisosteidae (Gars), Salmonidae (Salmons), Esocidae (Pikes), Cyprinidae (Minnows), Ariidae (Catfishes), Ictaluridae (Catfishes), Siluridae (Catfishes), Gadidae (Codfishes), Cyprinodontidae (Killifishes), Cottidae (Sculpins), Stichaeidae (Pricklebacks or Shannies).

(b) Epidemiology and Toxicity:

In ichthyootoxic fish, it is the reproductive organs and their products that are poisonous. The consumption of **roe** (fish eggs) sometimes creates health problems and cases of such poisonings are reported predominantly from freshwarer and marine species in Europe, Asia and North America *viz.,* **minnows** (Cyprinidae) of the genus *Cyprinus, Barbus, Tinca* and *Schizothorax*, as well as *Stichaeus* sp.

(Pricklebacks). The presence of toxins is mainly correlated with the breeding season. The identified toxic factor is ***dinogunellin*** (a lysophospholipid). Although such **toxic phospholipids** have been found in the roe of individual fish, it is still not certain whether the toxin is produced by the fish itself or ingested *via* food chain. In Japan, the roe of **Northern Blenny** or **Japanese prickleback** (*Stichaeus grigorjewi*) [Fig. 1.2a], has long been known to be toxic, causing a number of human poisonings. Recently, the **toxin** in the roe of a **marbled sculpin**, the **cabezon** (*Scorpaenichthys marmoratus*) [Fig. 1.2b], a well known ichthyootoxic species on the Pacific coast of North America, was found to be identical to *dinogunellin*. The toxin inhibits the growth of cells in tissue culture and produces **necrosis** in the liver and spleen.

Fig. 1.2: (a) Prickleback (*Stichaeus* sp.); (b) Marbled Sculpin, (*Scorpaenichthys marmoratus*).

(c) **Symptoms:**

Abdominal pain, nausea, vomiting, diarrhoea, headache, fever, bitter taste and dryness of mouth, intense thirst, a sensation of constriction in the chest, cold sweats, irregular pulse, low blood pressure, cyanosis, pupil dilation, chills, dysphagia and tinnitus are the common ailments associated with this kind of poisoning. In severe cases there may be muscle cramps, paralysis, coma and even death.

(d) **Treatment/Prophylaxis:**

Symptomatic. Although cooking is said to destroy most ichthyootoxins but this is not completely safe, since the poison in fish roe is heat resistant.

(iii) Ichthyohemotoxic: Fish with Poisonous Blood

(a) **Species/Group** (Table 1.1):

- **Osteichthyes; Anguilliformes:**
  Anguillidae (Fresh, brackish and marine eels), Muraenidae (Moray eels) [PLATE- I c], Congridae (Conger eels), Ophichthyidae (Snake eels).

(b) **Epidemiology and Toxicity:**

Freshwater eels (Anguillidae) spend their lives mostly in freshwater but migrate to sea for reproduction. **Moray**, **Conger** and **Snake eels** mostly live in coastal waters, on rocky coasts or in coral reefs. **Heat-labile toxins** are found in their blood. There is always a risk only if they are consumed raw or not cooked properly. The toxins are inactivated by boiling or frying as well as by the gastric juices.

The toxin derived from blood serum of an **Australian long-finned eel** or **marbled eel** (*Anguilla reinhardtii*) was used by **Charles Richet** in his Nobel Prize winning research about **'anaphylaxis'** (BOX 1.1). A proteinaceous toxin has been purified from the serum of **Japanese eel** (*Anguilla japonica*) and found lethal to both mice and crabs.

BOX: 1.1

CHARLES ROBERT RICHET

Charles Robert Richet (August 25, 1850 – December 4, 1935) was a French physiologist who worked on a variety of subjects *viz.*, neurochemistry, digestion, thermoregulation, breathing etc in homeothermic animals. He won the Nobel Prize for Physiology in recognition of his work on 'anaphylaxis' in 1913. The term anaphylaxis was coined to designate the sensitivity developed by an organism after it had been given a parenteral injection of a colloid or proteinaceous substance or a toxin. If two doses are given in one session or at a brief interval or even four or five days apart there are no toxic effects. There is an incubation period before the injected toxin has time to cause hypersensitivity in the organism. With regard to toxins, Richet called this hypersensitivity as anaphylaxis, in contrast to **phylaxis** or **prophylaxis**, meaning protection. Therefore, anaphylaxis is analogous in several ways to the phenomenon of immunity.

(c) **Symptoms:**

Drinking fresh, uncooked blood causes the symptoms like diarrhoea, bloody stools, nausea, vomiting, hypersalivation, skin eruptions, cyanosis, apathy, irregular pulse, weakness, paraesthesias, paralysis, respiratory distress and possibly death. Ocular contact invokes a severe burning sensation and redness of the conjunctiva, lacrimation and swelling of the eyelids.

(d) **Treatment/Prophylaxis:**

Symptomatic.

(iv) Scombroid or Histamine Poisoning

(a) **Species/Groups** (Table 1.1):

- **Osteichthyes:**

Clupeidae (Herrings, Sardines), Engraulidae (Anchovies), Scomberesocidae (Sauries), Pomatomidae (Bluefish), Carangidae (Jacks, Scads, Pompanos), Coryphaenidae (Dolphinfish), Scombridae (Mackerels, Tunas, Bonitos, Skipjacks, Albacore), Xiphiidae (Swordfish, Marlins).

(b) **Epidemiology and Toxicity:**

After ciguatera (see ahead), **scombrotoxism** is by far the most common form of fish poisoning. The **Scombroid poisoning** accounts for about 5% of food-borne poisonings and 38% of all seafood-associated outbreaks. All scombrotoxic fish are found in tropical to cold seas. In United States, especially in Hawaii, scombrotoxism is a common cause of seafood poisoning and the risk is often highest after ingestion of recreational catches. In countries like Denmark, France, Finland, New Zealand etc. the number with high outbreak rates ranges from 2 to 5/year/million people.

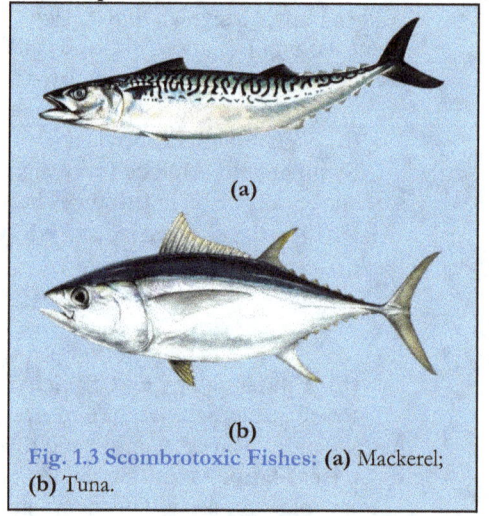

Fig. 1.3 Scombrotoxic Fishes: (a) Mackerel; (b) Tuna.

Unlike many other poisonous animals, the toxic substances in scombrotoxic fish do not accumulate *via* food chain. They, however, occur in the form of water soluble **histamine** and other products of decomposition *via* putrefactive processes in dead animals. Besides many other species, originally, **histamine poisoning** was found prominently in **mackerel** and **tuna** families, hence **scombroid poisoning** (Fig. 1.3). These long-distance, fast-swimming fish have dark red muscle tissue, rich in **histidine** amino acid. Under natural conditions, **histidine is converted into histamine** by decarboxylase enzymes, commonly found in the **halotolerant** (salt tolerant) or **halophilic** (salt loving) bacterial flora (*Clostridium, Klebsiella, Vibrio* spp. etc.). These bacteria flourish on the gills, external surfaces and in the gut of live, marine fish, without causing any harm to the fish. After death, the defense mechanisms of the fish no longer inhibit bacterial growth in muscle tissue and histamine-forming bacteria keep on growing, resulting in the production of **histamine**.

Further, because of some specific harvesting practices, such as long lining and gill-netting, death may occur hours before the fish is removed from water. Under such conditions, histamine formation can already be underway before the fish is brought on board. This condition further aggravates in **tuna** species that generate heat (hence, Tuna is often called 'warm-blooded'), resulting in internal temperatures exceeding environmental temperatures and creating conditions favourable for the growth of decarboxylase enzyme-forming bacteria. **Histamine** (= scombrotoxin) forming bacteria are capable of growing and producing histamine over a wide temperature range, being more rapid at about 20.0°C.

As the histamine-forming bacteria are facultative anaerobes, they also grow well in depleted oxygen conditions. As a result, **reduced oxygen packaging** (*e.g.*, vacuum packaging, modified atmosphere packaging and controlled atmosphere packaging) should not be viewed as inhibitory to histamine formation. Production of histamine is also facilitated at higher acidity levels (low *pH*). As a result, histamine formation is possible during various fish-preservation processes such as brining, salting, smoking, drying, fermentation and pickling until the product is fully shelf-stable.

Consumption of freshly caught fish or fish that have been continuously refrigerated (at <5°C), after being caught, is without risk. Evisceration and removal of the gills may reduce, but not eliminate, the number of histamine-forming bacteria. Although, packing of the visceral cavity with ice facilitates chilling of large fish, yet the internal muscle temperatures are not easily reduced. When done improperly, these steps may accelerate the process of histamine formation in the edible portions of fish by spreading bacteria from the visceral cavity to the flesh.

During storage or transport at higher temperatures, the decomposition process commences and the toxins cannot be destroyed even after frying, boiling or smoking. It is not possible to identify poisonous fish from external characteristics, but those that are of doubtful freshness or have a hot, **pepper-like after taste** should be avoided. In USA, the Food and Drug Administration (FDA) has set the **critical limit** of histamine as 50 *mg*/100 *g* tuna. In refrigerated and frozen animals the histamine content is generally < 5 *mg* /100 *g*.

(c) **Symptoms:**

Histamine-like reactions with nausea, vomiting, flushing of face, swelling of lips, urticaria and pruritus are the common symptoms which usually subside within 12 hours.

(d) **Treatment:**

Symptomatic.

Rapid chilling of scombrotoxin-forming fish immediately after death is the most important element in any strategy for preventing the formation of scombrotoxin (histamine), especially for fish that are exposed to warm waters or air and for tunas which generate heat in their tissues.

(v) Tetrodotoxic fish: Ttetrodotoxin-Poisoning: Poisoning by Puffers

(a) **Species/Groups** (Table 1.1; PLATE - II):

- **Osteichthyes, Tetraodontiformes:**

Tetraodontidae (Puffers, balloonfish, blowfish, bubblefish, globefish, swellfish, toadfish, toadies, honey toads and sea squab), Diodontidae (Porcupinefishes), Canthigasteridae (Sharp-nosed puffers), Molidae (Molas).

(b) **Epidemiology and Toxicity:**

The **skin** and some **visceral organs** of fishes belonging to Tetraodontidae are highly toxic due to the presence of **tetrodotoxin** (a sodium channel blocker and one of the **most neurotoxic** natural substances known; BOX 1.2) but nevertheless the flesh of some species is considered a delicacy in both Japan (as *fugu*) and Korea (as *bok*). There are several cases of deaths from *'fugu poisoning'* in Japan every year. Between 1886 and 1963, 10,745 cases were recorded, with a mortality rate of almost 60%. Although *fugu* caused about 632 deaths between 2003 and 2020, that does not stop Japanese from eating 10,000 tons of *fugu* each year.

BOX 1.2

THE 'TETRODOTOXIN' IN NATURE

Tetrodotoxin (TTX) is known from no fewer than five Phyla and about 20–30 species. It furnishes a good example of evolutionary convergence. Besides Puffers and their relatives, Tetraodotoxin has been found in Gobies (*Gobius criniger* and *Yongeichthys criniger*), marine snails (*Babylonia japonica*, *Nassarius* sp.), flatworms (*Planocera multitentaculata*), starfish (*Astropecten* sp.), crabs (*Zosimus* sp.) and blue-ringed octopus (*Hapalochlaena maculosa*). It has also been detected in animals outside the marine environment *e.g.*, in the spawn of the California newt (*Taricha torosa*) and in the gland secretions of atelopid frogs (*Atelopus* sp.). Recently, about 150 strains of bacteria have been discovered that can synthesise Tetrodotoxin, reaching Puffers (or other animals) *via* food chain.

Other regions of **tetrodotoxin-poisoning** include Australia, Bangladesh, Brazil, Cambodia, China, Egypt, Hong Kong, Indonesia, Israel, Italy, Lebanon, Madagascar, Malaysia, Mexico, Morocco, Oman, Phillipines, Singapore, South Korea, Thailand, Taiwan, Turkey, USA and Vietnam.

The first case of puffer poisoning was reported from **India** (Veraval, Gujarat) in May, 2020. The Andaman and Nicobar Islands; with extensive reef habitat for a large number of fish species of commercial, ornamental, poisonous and venomous importance; with Exclusive Economic Zone [EEZ] and a coastline of about 2000 kms; they are a repository of about 750 species of fish and out of them 200 species are considered to be poisonous and one of them is **puffer fish**.

THE 'PUFFERS' - *Fugu*

The **'Puffers'** or the **'*Fugu*'** (*e.g., Tetraodon, Takifugu* sp.) of Japan, include about 100 different species and are the best known of the toxic fish belonging to Tetraodontidae (max. size 90 cm) from marine and estuarine environment. *Tetraodon* and *Takifugu* sp. (PLATE II) are the second most poisonous vertebrates in the world [the first being a Golden Poison Frog, golden poison arrow frog or golden dart frog (*Phyllobates terribilis*), endemic to the Pacific coast of Columbia]. They are most diverse in the tropics and relatively uncommon in the temperate waters and completely absent from cold waters. The name, Tetraodontidae, refers to the **four large teeth**, fused to form upper and lower tooth plates used for crushing the shells of crustaceans and mollusks, their favourite food items.

Fig. 1.4: **Porcupinefish** (*Diodon* sp.).

Morphologically, they are quite similar to the closely related **porcupinefish** (*Diodon* sp.) [Fig. 1.4], having large conspicuous spines (unlike the small, almost sandpaper-like spines of Tetraodontidae). The **puffers** have the remarkable ability to inflate their body (PLATE-II) with air, thus exposing the long, sharp, toxic spines (modified scales), used for intimidating the predator. If however, an animal manages to prey upon the puffer, it is poisoned by the toxins in the spines or the toxin that is released from the organs of the puffer when it dies. If one is caught while fishing, it is recommended that thick gloves be worn to avoid poisoning and getting bitten when removing the hook.

THE '*FUGU*' FACTS

- **The Historical Facts:** The inhabitants of Japan have been using '*fugu*' as food for centuries. ***Fugu*** bones were found in several shell middens (an old dump for domestic waste), called ***kaizuka*** (a city located in Osaka), from the Jômon period, dating back to more than 2,300 years. **Tokugawa Shogunate** (also known as the Tokugawa bakufu and the Edo bakufu) was a feudal Japanese military government, ruling over Japan between 1600 and 1868), prohibited consumption of *fugu* in **Edo** (= 'bay-entrance' or 'estuary' and the older name of Tokyo). In western Japan, where the government's influence was weaker and *fugu* was easier to get, various cooking methods were developed to consume them safely. During **Meiji Era** (1867–1912), *fugu* was again banned. It is customary that the Emperor of Japan is forbidden to eat *fugu*, from safety point of view. ***Fugu*** was and is one of the favourite dishes in China, too, where its name was mentioned in the literature as early as *circa* 400 BCE. The most delicious *fugu* comes from the Yangtze river of China.

- **The *fugu* Symbols:** There is a bronze monument of fugu in front of the fish market at Shimonoseki (Japan). There are even fugu pictures on the city's manhole covers. A local tourist brochure reads, *"In the past, eating fugu was an adventure, risking one's life"*.

    The Etymology: In Japanese, the common name given to **Puffers** or **Blowfish** is '*Fugu*', having its origin from two '*kanji*' or '*Han characters*' or '*Chinese characters*' i.e., *fu* + *gu*. It is noteworthy that ***Kanji*** is symbolic or logographic, most common means of written communication in Chinese/Japanese language. '*fu* + *gu*' is actually a combination of two characters, respectively pronounced as '*he*' + '*tun*' and meant in Chinese as '**river**' + '**pig**' i.e., the common name of the '**puffer**' (= *fugu*) literally means in Japanese, a '**river pig**'. This is proven by the fact that the pufferfish which the Chinese used to eat centuries ago were found in large rivers like Yangtze River and when they were caught, they made a **wheezing sound** like the '*oinks of a pig*' (= grunting sound of a pig). Thus, the name '**river pig**' became a popularity.

    In the **Kansai** region (Osaka) of Japan, the slang word '*teppô*', (= rifle or gun) is also used for *fugu*. It comes from the expression '*teppo ni ataru*' (= to be shot). The word '*ataru*' (= to be poisoned or shot) means '**to suffer from food poisoning**'. In **Shimonoseki** region, the ancient pronunciation '*fuku*' (= good fortune, happiness, blessing, good luck) is more prevalent instead of the modern '*fugu*', possibly on account of the prized and deliciousness of the luxury fish.

    The fugu Flesh: As per National Geographic, "The meat has no fiber and is almost like gelatin. It is very light in taste. More like chicken than fish……..".

    Also known as the '**king of white fish**', *fugu* is rich in an organic acid (*2-aminoethanesulfonic acid*), the **Taurine** which is supposed to be quite vital for the functioning of the heart and brain. It helps support nerve growth. Having a property of lowering blood pressure and calming the nervous system, it also benefits people suffering from heart failure. While being low in fat (0.2 – 0.4%) and calories, fugu's flesh contains high level of collagen and protein (17.0 – 18.0%).

- **The Most Dangerous Part of the Fish** is the liver, which, according to Japanese people, is the tastiest meat.

- **The *fugu* Consumption Rate:** Japanese consume about 10,000 tons of *fugu* per year. Even though *fugu* is very popular in Osaka, Tokyo is the largest consumption center of *fugu*. **Shimonoseki**, on the southern tip of Honshu, is particularly famous for the best *fugu*.

- **The Fatality:** One gram of *fugu* poison (**tetrodotoxin**) is enough to kill 500 people and is about 500 to 1,000 times lethal than **potassium cyanide**.

- **The Toxin-laden Bacteria:** It is usually argued that *fugu* gets the poison after ingesting Tetrodotoxin-laden ***Vibrio*** bacteria, found in animals on which puffers feed *viz.*, starfish, worms and shellfish. Others opine that fugu's toxicity is due to poison glands beneath the skin.

- **The *fugu* Fishing Season:** The *fugu* fishing season lasts from October to February. It is a usual practice, that each morning at 4:30 am a lottery is drawn for deciding the location where the ships will fish. The vessels leave port before the dawn to catch puffers with long lines and returning to the port in the afternoon with about half the catch loaded onto trucks and transported live to Shimonoseki.

- **The '*fugu* Farming':** Scientists at Nagasaki University have reportedly succeeded in creating a nontoxic variety of **torafugu** or **tiger puffer** (*Takifugu rubripes*) [PLATE – II] by restricting the fish's diet. About 4,500 tons of *fugu* is produced annually. They are raised in large circular tanks and have their mouths sewn shut so that they do not attack one another. Thus, one of Japan's main **torafugu farms** is in Nagasaki Prefecture in northwest Kyushu Island where this rare delicacy is cultivated, by blending **Saikai mandarin orange** (a citrus fruit; native of south-eastern Asia and the Philippines) into their feed. It is believed that the inclusion of this citrus fruit in the **torafugu's** diet improves blood flow and results in firmer flesh. Due to the increase in demand for T. *rubripes* in Japan, the amount of cultured *torafugu* is increasing in Japan and in other countries like mainland China and Korea. Recently, it has been successfully cultured in Taipei and Ilan Prefecture (Northern Taiwan).

- **Trading Centres in Japan:** In Japan, the 'puffer's trade' is strictly regulated by local governments. Those, who do not abide by the ordinances on the sale in Tokyo, can be imprisoned for up to two years and fined ¥500,000. The government promotes exports of *fugu*, targeting wealthy people in China and other Asian countries. Haedomari, in Shimonoseki (Japan), is the sole market, specialized in *fugu* trade. The auction starts after 3:00 am. About 300 tons wild puffers are put on auction every year, with an average of about 1 ton/day. Quality fish, known as **beppin** or **beauties**, is sold mostly to high-end restaurants in Tokyo and Osaka. Puffers sold at the market are distributed as '*fugu* **from Shimonoseki**' and is regarded as a top quality fish both in terms of taste and safety.

- **The 'blowfish/puffer trade' in US and Europe:** As of 2003, only 17 restaurants in United States were licensed to serve *fugu*; 12 of those were in New York. Since then, some other American restaurants also offer *fugu*. The fugu is cleaned off the most toxic parts in Japan and freeze-flown to USA under license in customized, clear, plastic containers. *Fugu* chefs for US restaurants are trained under the same rigorous specifications as in Japan. Pufferfish, native to US waters, particularly the **Spheroides** sp., (PLATE – II) is also consumed as food, sometimes resulting in poisonings. Sale of **Spheroides** sp., is forbidden altogether in the European Union.

- **Top 'Puffer' exporting countries:** As of 2020, among the top 10 exporting countries of Pufferfish, China ranks first with its share of about 22.2% ($1.11B); Indonesia, US and India, respectively, ranking as second and third with their share of about 6.9% ($345.81M; for both Indoneasia and US) and 6.3% ($316.68M). Rest of the exporters are Myanmar (4.74%), Taiwan (3.5%), Senegal (2.33%), Vietnam (2.9%), South Korea (2.14%) and Japan (2.03%).

    Ordinance on *Fugu* preparations: In Japan, 22 different kinds of *fugu* have been approved by the Government for use in restaurants. The Tokyo metropolitan government enacted an ordinance on *fugu* in 1949, the nation's first licensing system for preparation of the fish, in the wake of a number of deaths from *fugu* poisoning during a period of food shortages after World War II. Under Tokyo's regulations, only the **licensed *fugu* chefs** are permitted to buy, process and sell fresh *fugu* and **migaki fugu** (= *fugu* that has its poisonous parts removed).

    However, very recently (2013), the stringent ordinance on the *fugu* trade has been eased out. Under the new rules set by the Tokyo metropolitan government, shops and restaurants, that do not have

---

BOX 1.3
SOME POPULAR *FUGU* DISHES

- **Fugu sashimi** or **Fugu Sashi** or **Tessa:** *fugu* cut into translucent slices (PLATE – II).
- **Milt:** The highly prized food item in Japan is the soft roe (*Shirako* or *Japanese Butterly Fish Semen*) available in departmental stores. It is quickly heated and served with ponzu (a citus-based sauce).
- **Fugu Kara-age:** The deep-fried *fugu*.
- **Hire-zake:** Baked dish out of dried fins and served in hot sake (an alcoholic beverage).
- **Stew** or **Fugu-chiri** or **Tetchiri**: The dish of simmered (boiled) vegetables and *fugu*.
- **Yubiki:** The skin, eaten as part of a salad.
- **Hirezake:** Cooked *fugu* tail in hot sake, traditionally sipped with a *fugu* dinner.
- **Ovary:** The ovary contains greater amounts of the lethal poison tetraodotoxin than other parts of the body. However, in Hakusan, Ishikawa, the toxin is eliminated in some local cuisine (Blowfish Ovaries Pickled in Rice-Bran Paste *via* preservation in salt and pickling in rice-bran paste).

certified *fugu* chefs, are not authorized to sell **migaki fugu**. This type of *fugu* is primarily served in the form of dishes like **'sashimi'** and **'hot pots'**. Also, the shops and restaurants without certified *fugu* chefs have to display a sign showing that they only serve **migaki fugu** and are required to report to the metropolitan government. They are directed to purchase certified **migaki fugu**, labeled with an inscription that the poisonous parts have been removed and to keep records about the place from where the fish was bought.

- **The Eating Season:** *Fugu* is considered a winter delicacy. Every February, people pray before a special shrine for a good puffer catch and fishermen usually send *fugu* as a gift to the Emperor. Besides the most prestigious edible and the most poisonous species, **Torafugu or tiger blowfish** (*Takifugu rubripes*), other preferred species are **Higanfugu** (*T. pardalis*), **Shôsaifugu** (*T. vermicularis* syn. *snyderi*) and **Mafugu** (*T. porphyreus*).

- **Fugu Restaurants in Japan:** Tokyo has around 800 certified *fugu* restaurants and there are about 3,800 throughout Japan. A famous restaurant specialized in *fugu* is **'Takefuku'** in the Ginza district (Tokyo). **Zuboraya** is another popular chain in Osaka.

- **The Trained Chefs:** There are about 80,000 *fugu* chefs in Osaka alone. To prepare *fugu*, a cook has to follow 30 prescribed steps. It takes about 11 years to become a full-fledged ***fugu*** **chef**. All cooks in Tokyo, that prepare *fugu*, are licensed. They have to go through a three years' apprenticeship under a master, take intensive courses, pass a written exam and show skill in preparing about a dozen types of *fugu* dishes. The written examination lasts for two hours. The next day, they are given a *fugu*, knife and twin pans, for practical examination. In a 20-minute test, all the poisonous parts are to be put in one pan and the edible parts in the other. Then, the parts are to be labeled with plastic tags (red for toxic, **black** for edible) and a meal to be prepared with proper garnishing and artful arrangement. The toughest part of the test is separating out ovaries, one of the deadliest parts, which look almost identical to the male's testes, which are a delicacy. If they are mixed, one is supposed to be declared failed in the test. Around 800 to 900 people take the test every year, with about two third of them being declared as passed.

- **The *fugu* Preparation Strategy:** To prepare fresh *fugu*, a chef has to take a live fish from a tank and to render it unconscious, knock it with a blow on the head with a mallet. After the poisonous parts have been removed with a *hocho knife*, the fish is cut into pieces and then washed under water to remove toxins and blood. The poisoned organs are placed in special bags to be kept under lock and key and disposed off like radioactive waste in a special incinerator.

- **The Recepies and Their Cost:** Depending upon restaurants, the *fugu* dining cost may range between ¥6,000 and ¥30,000 [43.40 – 217.0 US $; @ 100 Japanese Yen (¥) = 0.72215 US $]. Sometimes, each fish can sell for around ¥15,000. Obviously, the high costing of the fish is due to relative rarity of *fugu* and the intense training of a chef. Generally, a ***fugu*** **meal** sells between $40 and $100/person and typically has five courses with raw *fugu*, followed by fried fugu and stews, soups and broths made with parts such as skin and testes. **The thinly sliced raw flesh of the tiger *fugu* is supposed to be the best** (PLATE II). A plate of paper thin slices, arranged into patterns of flowers or birds, costs for over $200. Women like dishes with the skin because it contains large amounts of collagen, which is said to be good for skin. *Fugu* **testicles**, eaten before the mating season, are regarded as a special delicacy.

- **The 'Poisonous Delicacies':** The poisonous livers and intestines are made into a broth, the ***chiri*** *fugu*.

- **Zombies and Puffer:** In books *'The Serpent and the Rainbow'* (1985) and *'Passage to Darkness: The Ethnobiology of the Haitian *Zombie'* (1988), the Ethnobiologist Wade Davis argued that Zombies were created by giving 'Tetrodotoxin' taken out of the puffer. When given in a right dose, it makes the victim descend to near death and then awaken in a Zombie-like state.

***Zombie**: is an animated corpse raised by magical means, such as witchcraft or a mythological undead corporeal revenant created through the reanimation of a corpse. The term is often figuratively applied to describe a hypnotized person bereft of consciousness and self-awareness, yet ambulant and able to respond to surrounding stimuli. Since the late 19th century, **zombies** have acquired notable popularity, especially in North American and European folklore.

- **Murder by using *Fugu*:** In January 2011, a British businessman who died suddenly in Sierra Leone was suspected poisoned, using puffer-fish toxin.
- **'One Fish, Two Fish, Blowfish, Blue Fish'** is the eleventh episode of ***The Simpsons'*** (a satirical depiction of American life) second season. It was originally televised on the Fox network in the United States on January 24, 1991. The said episode was selected for release in a video collection of selected episodes, entitled *'The Last Temptation of Homer'*, released on November 9, 1998. In the said episode, Homer consumes a ***poisonous fugu*** fish at a *sushi* restaurant and is told that he has only 22 hrs left to live. He accepts his fate and makes a list of all the things he wants to do before he dies. The episode was written by Nell Scovell and directed by Wes Archer.
- A ***Rakugo*** [Japanese verbal entertainment of a gathering] or **Humorous Short Story:** A story narrates about three men who prepared a *fugu* stew but were not sure whether it was safe to eat. To test the stew, they gave some to a beggar. When it did not seem to do any harm, they ate the stew. Later, they met the beggar again and were delighted to see that he was still in good health. The beggar actually had hidden the stew instead of eating it. The three men were, thus, befooled by the wise beggar.
- **The *fugu* Toys:** Lanterns are often made from the bodies of preserved *fugu*, seen hanging outside the *fugu* restaurants, as children's toys, folk art or as souvenirs. *Fugu* skin is also made into everyday objects like wallets or water-proof boxes.

   A Poet's Sentiments: A **Haiku* poet Yosa Buson or Taniguchi Buson wrote:
   I cannot see her tonight,
   I have to give her up,
   So I will eat 'fugu'.
   [**Haiku* is a traditional Japanese three-line poem. It emphasizes simplicity, intensity and directness of expression].

THE TOXIN - TETRODOTOXIN (TTX)

- **HISTORY:**

    The first **Chinese Pharmacopoea** (about 2838 – 2698 BCE), **Pen-T'so Ching** (The Book of Herbs), reported therapeutic uses of 'puffer' eggs, supposed to be having 'mild toxicity' and tonic effects in cases of 'convulsive diseases'. Similarly, in **Pen-T'so Kang Mu** (Index Herbacea) by **Li Shih-Chen** (1596), **Ho-Tun** fish (= *Tetrodon* in Chinese) was the recognized 'puffer', both for its toxicity as well as usefulness as a tonic in right dose. Increased toxicity of **Ho-Tun** was noted in fish caught from marine environment (rather than river) after March. It was recognized that the most poisonous parts were the liver and eggs, but that toxicity could be reduced by soaking the eggs in water, proving thereby that tetrodotoxin is slightly water-soluble and soluble at 1.0 *mg/ml* in slightly acidic solutions. The German physician **Engelbert Kaempfer** (1727) in his book on *'A History of Japan'*, described the toxicity of the fish, to the extent that it could be used in suicidal attempts and that the Emperor specifically instructed the soldiers not to eat the 'puffers'.

    The first recorded case of **TTX-poisoning**, affecting Westerners, is from the official written records of **Captain James Cook** *vide* 7th September, 1774, when he recorded about his crew eating some local 'puffer', then feeding the remains to the pigs kept on board. The crew experienced numbness and breathlessness, while all the pigs were found dead the next morning. Supposedly, the crew survived a mild dose of tetrodotoxin, while the pigs which were fed on the toxin-laden 'puffer' body parts, were fatally poisoned.

    Japanese people used to eat *fugu* for quite long time, but the data about 'globefish' or 'puffer' or '*fugu*' poisoning were accumulated with Japanese Government since 1879. Chemical investigations into pufferfish toxin were started at Tokyo Imperial University by Juntaro Takahashi (1856-1920) and Kichindo Inoko (1866-1893). **Dr. Yoshizumi Tahara**, the Director of the Tokyo Institute of Hygienic Sciences, was curious about these studies and as such he intended to initiate conducting a chemical investigation on the toxic substance of globefish in 1884 and got success in isolating the toxin from pufferfish ovaries and partially purifying it. In 1909, Dr. Tahara confirmed that globefish contains only one toxic substance, named by him as **Tetrodotoxin** (tetrodo + toxin) after the *Tetraodontidae* family.

- **CHEMISTRY:**
    TTX ($C_{11}H_{17}O_8N_3$) is both water-soluble and heat stable. In crystalline form, it is a weak, basic, colourless substance. At least 30 structural analogues of TTX have been identified, with varying degree of toxicity. Depending on structure, these are classified into three groups *viz.*, ***hemilactal***, ***lactone*** and ***4, 9-anhydro*** types, altogether referred to as Tetrodotoxins (TTXs).
- **SOURCE:**
    TTXs are found in various, taxonomically diverse groups of both terrestrial and aquatic animals [see BOX 1.2]. In puffers, TTX is concentrated in the ovary and liver but other organs including skin, intestine and muscle also contain TTX.

    Most observations suggest that TTXs are either produced by **symbiotic bacteria** (endogenous route) or that they are exogenously accumulated through the diet. At least 150 bacterial strains have been found as TTX-producers, with the ***Vibrio*** spp. being the major representative.

    Besides, *Peudomonas*, *Aeromonas*, *Alteromonas*, *Nocardiopsis*, *Bacillus*, *Shewanella*, *Roseobacter* spp.etc. have also been found in the subcutaneous mucus, ovaries and the gastrointestinal tract of a number of aquatic species. Evidences are also there to associate some dinoflagellate blooms (*e.g., Alexandrium tamarense* or *Prorocentrum cordatum*) with TTX.
- **TOXICITY and MECHANISM OF ACTION OF TTX:**
    Most cases of Puffer-poisoning are caused by ingesting toxic livers. TTX is highly toxic to mammals with an LD_{50} in the order of 10.0 *μg/kg*. TTX is potentially **neurotoxic**, with a unique property of blocking the passage of sodium ions through the cell membranes of nerve cells, without changing the resting membrane potentials. The voltage-gated sodium channels are essential ion channels for resting potential and neuronal excitability, which are crucial for the generation and propagation of action potentials in neurons. Thus, there is a primary blockade of the brainstem, somatic motor, sensory and autonomic nerves. It ultimately leads to paralysis of nerves and muscles. Due to its blocking of specific Na^+ ion channels, called ***TTX-resistant sodium channels***, TTX has been found to be used as a medical tool to express analgesic and local anesthetic properties.

> ***TTX-resistant sodium channels:*** Investigations on the **medical use** of **TTX** have revealed its **analgesic** and local **anesthetic properties**, due to its blocking of **very specific Na^+** ion channels. In early 1980s, specific **TTX-resistant sodium channels** had been reported for the first time. Drugs that block TTX-resistant sodium channels without affecting TTX-sensitive sodium channels are found to serve as useful **painkillers**. In early 20th century, TTX was found to work **as an analgesic agent** for the treatment of neuropathic and rheumatic pain in Japan. TTX has also been used as an analgesic agent in advanced **cancer patients** in China. A Canadian Pharmaceutical Company (WEX Pharmaceuticals, Inc., Vancouver, Canada) developed a drug, containing TTX for subcutaneous injection, as an analgesic in advanced cancer patients to reduce the intense pain as an alternative to narcotics and opioid pain medication and the treatment of opiate addiction. **TTX-resistant sodium channels** are known to exist in cardiac muscles, dorsal root ganglion cells and **'C' fibers** of dorsal root ganglion cells that convey pain sensation to the brain. It should be noted, however, that Na^+ channels present in TTX-carrying animals such as pufferfish and some types of shellfish, frogs, salamanders, octopuses, etc., are resistant to TTX.

(a) Symptoms:
A low dose of tetrodotoxin produces **paresthesias** (tingling/prickling/itching sensations and numbness) around mouth, fingers and toes within 10 to 45 minutes after ingestion of the fish. Higher doses produce nausea, vomiting, respiratory failure, difficulty in walking, extensive paralysis and death within the first 6 hours or within 24 hours. As little as 1.0 – 4.0 *mg* of the toxin can kill an adult. As per Japanese statistics, the fatality rate is about 61.5%. Based on symptoms, *Fugu*-poisoning is often categorized into **4 Grades**, of which **Grade 3** and **Grade 4** are considered as the **most severe** ones:

GRADE 1: Neuromuscular symptoms *viz.*, perioral paresthesia, headache, sweating and pupil constriction and mild gastrointestinal symptoms *e.g.*, nausea, vomiting, excessive salivation, diarrhea and abdominal pain.

GRADE 2: Paresthesia extending to the trunk and limbs, early motor paralysis and lack of coordination.

GRADE 3: Worsening of neuromuscular symptoms *viz.,* dysarthria, dysphagia, sleepiness, ataxia, feeling of floating, cranial nerve palsy, muscle tremor, heart/lung symptoms (low blood pressure or hypertension, vasomotor dysfunction, cardiac arrhythmias, cyanosis, pallor, difficulty breathing, etc), skin disorders (bruises, exfoliative dermatitis, blisters), hypotension and aphasia.

GRADE 4: Delirium, respiratory paralysis, severe hypotension and arrhythmia.

(b) Treatment:
Symptomatic.

There is no known antidote but it is claimed that a monoclonal antibody against tetrodotoxin (anti-tetrodotoxin) has been developed by *United States Army Medical Research Institute of Infectious Diseases* (**USAMRIID**), but without much authentication of its efficacy. For suspected tetrodotoxin-poisoning, patients are required to be keenly observed in intensive care unit (ICU) for 24 hours, because some patients show a delayed onset of symptoms after up to 20 hours. The usual course of treatment is *via* **respiratory support**, until the tetrodotoxin is excreted in urine. Medicos often administer *Neostigmine* (or *Bloxiverz*) to treat acute respiratory failure from tetrodotoxin poisoning. If a patient is reported of the poisoning within 1.0 hour of ingestion, gastric decontamination is done through **gastric lavage**, especially with 2% sodium bicarbonate solution, followed by activated charcoal, since TTX is less stable in an alkaline medium. **Hemodialysis** is often useful in patients with renal disease.

(vi) Clupeotoxic Fish: Poisonous Herrings, Sardines and Anchovies

(a) Species/Group (Table 1.1):
- **Osteichthyes, Clupeiformes:**
  Elopidae (Ladyfishes), Albulidae (Bonefishes), Clupeidae (Herrings and Sardines) and Engraulidae (Anchovies).

(b) Epidemiology and Toxicity:

All the clupeid fishes are economically important edible fish. Their body is generally elongated and slender. Grey-green colouring on the back and shiny silver on the belly and sides is diagnostic to them. Most swim in huge schools in open waters and primarily feed on plankton. **Clupeotoxic fish-poisoning** is rare and resembles ciguatera (see ahead), but is very rapid in action. It is most prevalent during the warmest months, when the consumption of herrings, sardines and anchovies is generally avoided.

Sporadic cases are known from island regions of the tropical Atlantic and Pacific as well as the Caribbean. It is suggested that ***Palytoxin*** (PTX or PLTX) is the main cause of **Clupeotoxism** (BOX 1.4), as after consumption of small quantities of a **sardine** (*Herklotsichthys quadrimaculatus*) in Madagaskar, *palytoxin* was found in the tissues of the victim. Toxic fishes leave a metallic or bitter taste in the mouth when consumed.

PTX is a **vasoconstrictor** and one of the most poisonous non-protein toxins known, second only to **maitotoxin** (BOX 1.4) in terms of toxicity. It is a long carbon chain, **polyhydroxylated** (with as many as 40 hydroxyl groups) and partially unsaturated compound (8 double bonds). As per origin, it is found to be synthesized by **zoanthid corals** (*Palythoa* sp.) [BOX 1.4]. It is also observed that the toxin gets accumulated *via* food chain *i.e.,* ingesting highly **toxic,** temperate/tropical/subtropical, benthic, harmful marine algal blooms of *Ostreopsis* spp. (PLATE III a).

(a) Symptoms:

Nausea, vomiting, abdominal pain, diarrhea, sensory disturbance, paralyses, muscular pain, arrhythmia, arterial hypotension etc. are common symptoms. Clupeotoxism is said to have a high mortality rate and death in less than 15 minutes. Fish processing by boiling, salting or drying does not decontaminate them.

(b) Treatment/Prophylaxis:

Symptomatic and supportive *e.g.,* intravenous transfusion of 0.9% NaCl (@ 10-20 *ml/kg*), sometimes along with **dopamine** or **norepinephrine**, for treating severe hypotension; intravenous transfusion of **benzodiazepines** and **barbiturates** for persisting seizures; use of **analgesics** in case of headache; use of **antihistamines** in cases of pruritus and urticarial dermatitis.

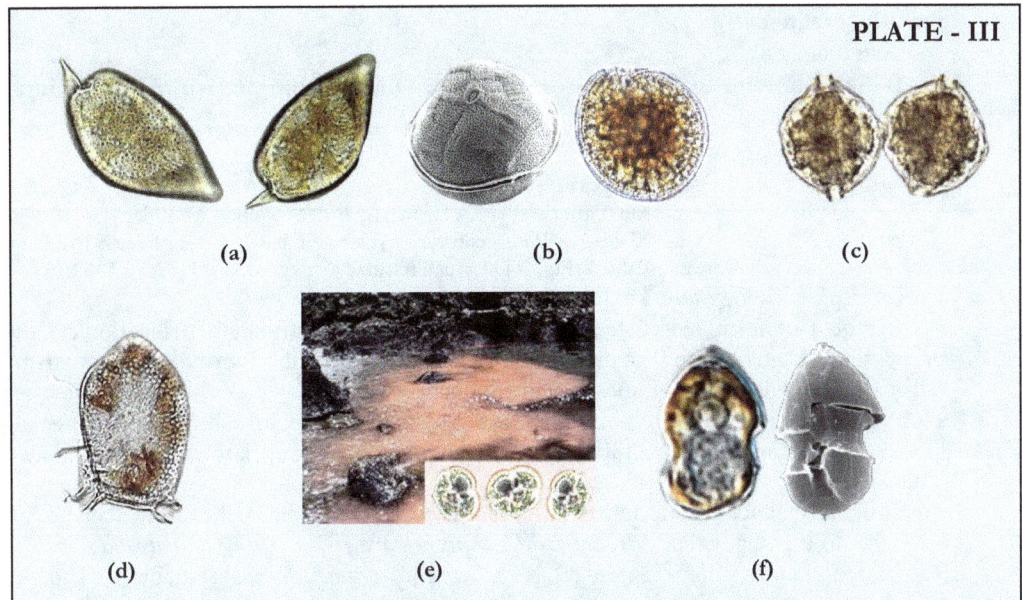

Harmful bloom-forming dinoflagellate(s)/diatom(s): **(a)** Palytoxin-producing, *Ostreopsis* sp; **(b)** Ciguatoxin-producing, *Gambierdiscus toxicus*; **(c)** PSP toxin-producing *Alexandrium* sp.; **(d)** DSP toxin-producing, *Dinophysis* sp.; **(e)** NSP toxin-producing, *Karenia brevis*; **(f)** AZP toxin-producing, *Azadinium spinosum*.

(vii) Gempylotoxic Fish or Gempylid Diarrhea or Escolar Diarrhea

(a) **Species/Group** (Table 1.1):
- **Osteichthyes, Perciformes:**
  Gempylidae (*Ruvettus pretiosus,* Oilfish, Snake Mackerel, Escolar, Rudderfish or Rutterfish)

(b) **Epidemiology and Toxicity:**

Two fishes belonging to Snake Mackerel Family, Gempylidae *viz., Ruvettus pretiosus* (Oilfish or Cocco) and *Lepidocybium flavobrunneum* (Escolar) (Fig. 1.5) often found in tropical and temperate oceans at depths between 100 – 800 *m*, are sold under the category of **'butterfish'** and contain a strong purgative oil. When consumed, this oily substance causes diarrhea known as **Gempylid Fish Poisoning, Gempylotoxism** or *****Keriorrhea**. The toxin involved, **Gempylotoxin** is a complex of **wax esters** (C_{32}, C_{34}, C_{36} and C_{38} fatty acid esters), forming a considerable amount of the lipid present in the fish (14 –25% by weight). Ingestion of these wax ester-containing fish in large amounts, coupled with their indigestibility and low melting point, results in diarrhea.

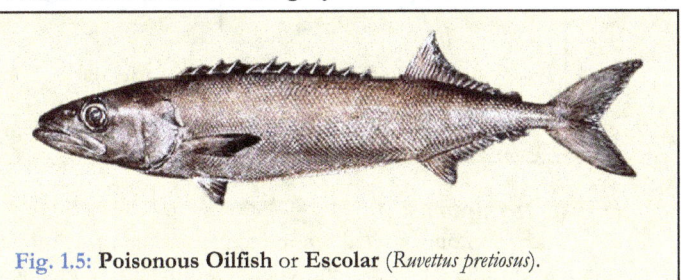

Fig. 1.5: **Poisonous Oilfish** or **Escolar** (*Ruvettus pretiosus*).

*****Keriorrhea** = an oily, orange-coloured bowel movement that occurs when a person consumes indigestible wax esters.

(c) Symptoms:
Loose motions (Diarrhea).
(d) Treatment:
Symptomatic.

(viii) Hallucinatory fish poisoning (Ichthyoallyeinotoxic fish) or hallucinogenic fish poisoning

(a) Species/Group:
- **Osteichthyes: Perciformes:**
  Serranidae (Groupers) [PLATE I f], Sparidae (Saupe), Mullidae (Goatfishes), Kyphosidae (Rudderfishes), Pomacentridae (Damselfishes), Mugilidae (Mullets), Siganidae (Rabbitfishes) [Fig. 2.11], Acanthuridae (Surgeonfishes) [Fig. 2.12, 2.13].

(b) Epidemiology and Toxicity:

Certain Perciform reef fish (as listed above); found distributed in the tropical Indian, Pacific Oceans and Mediterranean Sea, are associated with what is called **Ichthyoallyeinotoxism** (Gk. *ichthys* = fish + *aluein* = to be out of oneself, to hallucinate + *toxikon* = venom). It is classified as a **variant of ichthyosarcotoxism** *i.e.,* food poisoning caused by the ingestion of fish flesh. This type of poisoning is sporadic, uncommon and absolutely unpredictable. Outbreaks are known to occur in Hawaii and Fiji Islands.

This rare hallucinogenic intoxication is due to consumption of head or other body parts of fish belonging to a group, called *'dream fish'* or *'nightmare fish'* e.g., *Kyphosus vaigiensis* (Kyphosidae) [Fig. 1.6 a], *Sarpa salpa* (Sparidae) [Fig. 1.6 b], *Siganus spinus* (Siganidae) and *Mulloidichthys samoensis* (= *flavolineatus*) (Mullidae) [Fig. 1.6 c]. Reportedly, during Roman Empire *Sarpa salpa* was used for ceremonial purposes in Polynesia and for recreational purposes in countries surrounding the Mediterranean Sea. Because of their **psychoactive effects**, such fish are often referred to as **'psychoactive fauna'**.

As all ichthyoallyeinotoxic fishes are algal grazers, it is has been suggested that they derive their hallucinogenic properties from **alkaloids** of the **indole group**. The chemical structure displays similarities to **LSD** (*Lysergic Acid Diethylamide*) and occur naturally in some algae and phytoplankton. It has also been suggested that ichthyoallyeinotoxism is mediated by the presence of **Di-Methyl Tryptamine** (DMT), also a hallucinogen, even more potent than the indoles.

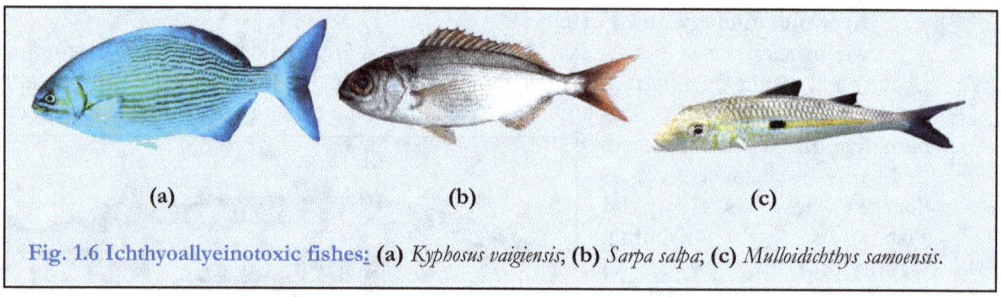

Fig. 1.6 Ichthyoallyeinotoxic fishes: (a) *Kyphosus vaigiensis*; (b) *Sarpa salpa*; (c) *Mulloidichthys samoensis*.

(c) Symptoms:

The symptoms typically commence within a few minutes to 2 *hrs* after consuming the fish, lasting no longer than 36 *hrs*. Ichthyoallyeinotoxism is characterized by **malaise** and a variety of **CNS disturbances** such as nausea, vertigo, disturbances in motor coordination, vivid visual and /or auditory hallucinations, nightmares and sometimes delirium. Sensation of tight constriction around the chest, terror, itching, burning of the throat, muscular weakness, rarely abdominal distress etc. are among other symptoms.

(d) Treatment:
Symptomatic

(ix) Ciguatoxic Fishes or Ciguatera Syndrome or Ciguatera Fish Poisoning (CFP)
(a) Species/Group:
- **Osteichthyes:**
    Anguilliformes:
        Muraenidae (Moray eels) [PLATE I c].
    Perciformes:
        Acanthuridae (Surgeonfishes) [Fig. 2.12, 2.13], Carangidae (Kingfishes), Labridae (Wrasses), Lutjanidae (Snappers), Scaridae (Parrotfishes), Serranidae (Rockcods and Seabasses, Groupers) [PLATE I e,f], Scombridae (Tunas, Mackerels and Bonitos) [Fig. 1.3], Sphyraenidae (Barracudas).

(b) Epidemiology and Toxicity:

Worldwide, **Ciguatera fish poisoning** (CFP) is the most common type of seafood-borne illness, caused by consumption of fish that have accumulated lipid-soluble **ciguatoxins** (CTXs). More than 400 herbivorous, carnivorous and omnivorous bony fish have been found associated with it, specially the larger carnivores like **Moray eels, Carrangids, Snappers, Groupers, Spanish Mackerels** *etc*.

Regular outbreaks are reported in the **primary endemic areas** like the Caribbean, the South Pacific Islands and Hawaii in the North Pacific, as well as the northeast coast of Australia (Queensland, Northern Territory). It has now spread widely to the Eastern Atlantic, Western Gulf of Mexico and Eastern Mediterranean. In the Pacific, the severity and incidence of poisoning increases from west to east. Isolated outbreaks occur with increasing frequency in temperate regions of Europe and North America. Brisk travelling between temperate countries and endemic areas and import of susceptible fish has led to spread of **CFP** into regions where it has rarely been encountered.

As many as 50,000 cases of **CFP** are reported worldwide annually and has been found to have a considerable impact on the Caribbean islands, Florida, Pacific Islands and Australia, because their economy is based mainly on fish trade and tourism. In **primary endemic areas** (the Caribbean and South Pacific Islands), the incidence is between 50 and 500 cases/10,000 people. Therefore, it is often categorized as a **pandemic,** impacting the most in the tropical and subtropical regions. As per a report by the **National Center for Toxicology**, **Cuba** (CENATOX), there had been about 300 cases each year, between 1989 and 1996. From 1992 to 1996, the highest percentage of **CFP** occurred between May and September, with a peak in June 1994, when 89 cases were reported.

ETYMOLOGY

Ciguatera [American Spanish *ciguato* = one poisoned with ciguatoxin from *cigua, sigua* snail] is regarded an **'umbrella term'** for ichthysarcotoxism. The term had its origin in 18th century from the aboriginal word **Cigua** or **Sigua**, the Spanish trivial name for a **Gastropod Mollusc** (Family: Tegulidae), the **West Indian top shell** or **Magpie shell** (*Cittarium* (*Livonia*) *pica* or *Turbo pica*) [Fig. 1.7]; also called *'siwa'* in the English-speaking Caribbean. This mollusc, common along the rocky coasts, is consumed raw and marinated in lemon juice. It is the third most common marine invertebrate eaten in West Indies and notoriously reputed for causing indigestion. Symptoms related to this snail are similar to those of Ciguatera poisoning and hence the Spanish conquerors expanded the term

Fig. 1.7 West Indian Top Shell: *Cittarium* (*Livonia*) *pica*

and used it to refer to this Gastropod and its consumption. Finally, the word Ciguatera was used in 1787 by **Don Antonio Parra** of Cuba (Havana) and then the Cuban naturalist **Felipe Poey** used it to describe similar cases. **CFP** is sometimes called *'Ciguatoxinosis'*, since ciguatera denotes the poisonous snail and not the toxin (Ciguatoxin). Later on, the term *'cigua'* was somehow shifted to an intoxication caused by the ingestion of coral reef fishes.

HISTORY

Early references to ciguatera-like illnesses are found in the 1555 Chronicle of West Indies by **Pedro Martyr *de* Anglería**, but earliest references include those of ancient Greek poet **Homer's Odyssey** (800 BCE) and an outbreak in **China** way back in 600 BCE. In the times of **Alexander the Great** (323-356 BCE), soldiers were prohibited from consuming fish to avoid illness during the conquests. It was reported from the islands of Indian Ocean by **Admiral Wolphart Harmansen** in 1601 and in various archipelagos of the South Pacific Ocean by the Portuguese navigator **Pedro Fernandez de Quiros** in 1606. In July 1774, members of British explorer, **Captain James Cook's** (famous for his three voyages between 1768 and 1779 in the Pacific Ocean) crew suffered ciguatera poisoning after eating fresh fish (a bream or snapper; *Sparus* sp.) caught off the coast of Vanuatu (South Pacific).

For **Cuba** (West Indies), Spanish sailors made the information about CFP available, but it was published for the first time in Havana in 1787 by a Portuguese, **Antonio Parras**, who described how his family suffered from seafood poisoning.

THE TOXIN AND ITS SOURCE

John E. Randall (1959) hypothesized that Ciguatera-causing toxin or **Ciguatoxin** is introduced into the food chain by herbivorous fish which subsist on toxic microalgae and are, in turn, consumed by large predatory fish. The identification and isolation of ciguatoxin was done by **P.S. Scheuer** (1967) and colleagues. **T. Yamamoto** and his co-workers (1977) ultimately discovered that it is a kind of **benthic dinoflagellate** (*Gambierdiscus toxicus*) [PLATE III b], involved in the production of ciguatoxin. The said dinoglagellate was found close to the Gambier Islands, on the surrounding fringing reef, in French Polynesia. **M. Murata** (1989) and his colleagues derived the structure of a major ciguatoxin and its precursor in *G. toxicus*. Other dinoflagellates *viz., Prorocentrum* spp., *Gymnodinium sangieneum* (Fig. 3.1 a, d) and *Gonyaulax* (= *Lingulodinium*) *polyedra* are also found to produce toxins associated with ciguatera poisoning.

Blooming of *Gambierdiscus toxicus*:

There is an increasing evidence that the factors which lead to a massive increase in dinoflagellates, and thus toxin production, are linked to destructive influences on coral reefs *e.g.,* hurricanes or heavy rainfall, construction of harbours, military operations, discharge of industrial sewage etc. The sediments and dead coral reefs offer a good substrate for fast-growing algae, which in turn attract epiphytic *Gambierdiscus toxicus*. Such epiphytic association is often observed with bushy, red, brown and green seaweeds.

Maitotoxins and Ciguatoxin:

The dead coral and marine algae thriving in tropical and subtropical reef systems are taken as food by herbivorous fish, accumulating and concentrating the toxins produced by *G. toxicus*. The said dinoflagellate is found to be the source of two types of toxins, *i.e.,* the water-soluble **Maitotoxins** (MTXs) and the fat-soluble **Ciguatoxins** (CTXs). **MTXs** are principally found in the gut of herbivorous fishes and have no proven role in CFP. On the other hand, **CTXs** are found in the liver, muscles, skin, bones and roe of large carnivorous fishes and are regarded as the principal cause of CFP.

MTX was named after a ciguatoxic fish *Ctenochaetus striatus* (Acanthuridae), called **'maito'** in Tahiti (the largest Southern Pacific island in French Polynesia) but later on it was shown that **MTX** is actually produced by *G. toxicus*. At molecular level, **MTX** resembles **a long fatty acid chain** and is one of the **largest** and **most complex** non-protein, non-polysaccharide molecule produced by any organism (comprising 32 ether rings, 22 methyls, 28 hydroxyls and 2 sulfuric acid esters).

CTXs are **polyethers** consisting of 13 to 14 rings fused together by ether linkages into a most rigid ladder-like structure. They are relatively heat-stable and remain toxic even after cooking and exposure to mild acidic and basic conditions. In the body of a fish, more potent oxidized forms of CTXs develop after biotransformation of the **precursor gambiertoxin-4B** (P-GTX-4B) derived from the dinoflagellates. The metabolic modification of dinoflagellate toxins in fish produces a large number of structurally related **CTX congeners**. To date, 47 CTXs have been identified, however, the structure of CTXs varies according to geographic distribution and as such; they are classified as **Pacific Ocean** (P-CTX), **Caribbean Sea** (C-CTX) and **Indian Ocean ciguatoxins** (I-CTX).

> **BOX 1.4**
> **PALYTOXIN and SCARITOXIN**
> Two **variants of Ciguatoxin** *viz.*, **Palytoxin (PTX)** and **Scaritoxin** are also known. **PTX** was discovered from fishes like *Decapterus macrosoma* (Scombridae), *Ypsiscarus ovifrons* (Scaridae), *Melichthys vidua* (Balistidae) and a crab (*Demania reynaudii*). It was first isolated from a sea weed-like Hawaiian Zooanthid **soft coral** (*Palythoa* sp.) popularly called, 'Limu make o hana' (= Seaweed of Death from Hana). Zoanthids (Anthozoa, Hexacorallia) are colonial anemones found to contain a large, complex, water-soluble **phytotoxin** with a long **polyhydroxylated** (= *Polyol*) and partially unsaturated aliphatic backbone ($C_{129}H_{223}N_3O_{54}$), called **PTX**. Being an intense **vasoconstrictor**, it is considered to be one of the most toxic **non-peptide** substances known, second only to **MTX** in terms of toxicity.
>
> An epidemiological survey of ciguatera intoxication at the Gambier Islands revealed that the most frequently implicated fish were **Parrotfish** (*Scarus gibus*; Scaridae). Affected persons, in addition to the conventional immediate ciguatera symptoms, suffered from disturbed equilibrium, locomotor difficulties and kinetic tremors. This new toxin from the Parrotfishes was called **Scaritoxin** ($C_{60}H_{84}O_{16}$).

The most studied variants are from the Pacific *e.g.*, **P-CTX-1** ($C_{60}H_{86}O_{19}$), **P-CTX-2** ($C_{60}H_{86}O_{18}$) and **P-CTX-3** ($C_{60}H_{86}O_{18}$) but **CTX-1** is the **major toxin** found in carnivorous fish, posing health risks among the consumers. It is worth mentioning that, during the biotransformation of **P-GTX-4B** to **P-CTX-1** there is a ten-fold increase in potency. As referred above, P-CTX-1 is regarded as the most potent toxin and the recommended safety limit for CTXs in fish for human consumption has been set at 0.01 *ng* P-CTX-1 toxin equivalent/*g* fish tissue by both the **European Food Safety Authority** (EFSA) and **United States Food and Drug Administration** (US FDA).

Two more variants of Ciguatoxin *viz.*, **Palytoxin** (PTX) and **Scaritoxin** are also identified (BOX 1.4).

*IDENTIFICATION OF CIGUATOXIC FISH

Globally, CFP has become problematic for the communities subsisting on seafood. Therefore, identification and quantification of CTXs has been a challenging task (even for laboratories), due to quite low concentration of CTX in fish flesh and lack of reference materials and standards for all CTXs. Therefore, currently, there is no routine, rapid, reliable and cost-effective test that can detect CTXs on-site or prior to consumption. However, there had been various *in vivo* whole-animal detection **indigenous methods** which are now in the process of being replaced by more sensitive *in vitro* assays *viz.*, **Receptor-Binding Assays** (RBAs), **Cell-Based Assays** (CBAs), **Enzyme-Linked Immunosorbent Assays** (ELISA), **Capillary Electrophoresis (CE)-Based Immunoassays**, **Electrochemical Immunosensors** (ECS) and **Liquid Chromatography Tandem Mass Spectrometry** (LC-MS/MS).

As far as **Indigenous Methods** are concerned, island communities practiced them over centuries, *viz.*, rubbing a small piece of liver on the mouth or skin and then testing for itchiness; cooking fish with a silver coin or copper wire and assessing discolouration; observing the colour of fish gallbladder; examining food avoidance by ants and flies; feeding dogs, cats or pigs with suspected fish and observing sickness or fatality on the animals and bleeding and *rigor mortis* tests. In a bleeding test, a fish was considered toxic if haemorrhagic symptoms are evident after an incision is made on the tail of the dead fish. In the ***rigor mortis*** **test**, a fish was considered to be toxic, if its flesh remained flaccid even an hour after death. All such practices were rejected and found invalid by the scientific community.

Although significant advances have been made in '*in vitro* ciguatera assays'; an ideal, simple, cheap, rapid, reliable, highly sensitive, quantitative assay; providing specific toxin profiles and not requiring trained operators and specialized equipment; is not available till date.

In general, **Antibody-based Immunoassay methods**, are fast and easy to use, the sole exception being the **Radioimmunoassay** (RIA). Production of antibodies is one of the key issues and constraints of immunoassays. In all cases, an antibody is labeled/marked to detect target-antibody interaction. This label can be a radioisotope, enzyme or a fluorescent probe. To assess the presence of CTXs in fish tissues directly, the **first RIA** was developed by **Dr. Y. Hokama** (Department of Pathology, John A. Burns School of Medicine, University of Hawaii, Honolulu, USA) in 1977.

[***NOTE:** for more details please refer to, '**Advances in Detecting Ciguatoxins in Fish**', *Toxins* (Basel): 2020: **12(8)**: 494: *https://www.ncbi.nlm.nih.gov* ›]

MECHANISM OF TOXIN ACTION

Levels of CTXs exhibit risk to human health at concentrations higher than 0.022–0.1 *ng/g* fish flesh. The concentration of toxins in fish liver is about 10–50 times higher than in muscle tissue, thereby CFP is becoming more problematic in communities consuming fish viscera. Analysis of ciguatoxins in blood samples suggests that the toxin is stored in adipose tissue.

CTXs are potential **sodium-channel activator neurotoxins**, that pose risks to human health at very low concentrations (> 0.01 *ng/g* of fish flesh in the case of the most potent **P-CTX-1**). They do not cross the **Blood Brain Barrier** (BBB), hence, solely affect the **Peripheral Nervous System** (PNS) by lowering the threshold for opening Na^+ channels at nerve synapses. Opening of Na^+ channels results into depolarization, sequentially leading to paralysis, contraction of heart and altering the senses for heat and cold. While **CTXs** act on Na^+ channels of nerves and muscles, **MTX** activates Ca^{2+} permeable, non-selective cation channels, leading to an increase in levels of cytosolic Ca^{2+} ions. It is supposed that MTX leads to the creation of pores on these ion channels.

Ultimately, a cell death cascade is activated, resulting in cell lysis. Furthermore, as a consequence of an influx of Ca^{2+}, MTXs trigger hormone and neurotransmitter secretion and phosphoinositide breakdown and activation of Ca^{2+} channels due to membrane depolarization. No specific blocker has been identified for this MTX-induced channel. MTXs get accumulated in the viscera of herbivorous fish, but obviously are not accumulated at sufficiently high doses in carnivorous fish to cause problems on consumption. Poisoning with **scaritoxin** is not well described.

(c) Symptoms:

A combination of a few to more than 30 gastrointestinal, neurological and/or generalized disorders has been reported. **Gastrointestinal symptoms** are a dominant feature in the Caribbean while **neurological symptoms** predominate in the Pacific Ocean region. In addition to the typical symptoms of ciguatera, a mixture of symptoms *viz.,* lack of coordination, loss of equilibrium, hallucinations, mental depression, nightmares *etc* are found prevalent in the Indian Oceanic region.

Since the analysis of ciguatoxins in blood samples suggests that the toxin is stored in adipose tissue, the symptoms may recur during periods of stress *e.g.,* exercise, weight loss or excessive alcohol consumption. After consumption of CTX contaminated fish, the onset of the first symptoms begin after 30 minutes for severe intoxications, while in milder cases onset may be delayed for up to 24 - 48 *h*. Ciguatera symptoms typically last for several weeks to several months. In less than 5% cases, certain symptoms may persist for a number of years, in extreme cases as long as 20 years, often leading to long-term disability.

Gastrointestinal symptoms, involving vomiting, diarrhoea, nausea and abdominal pain (>50% of cases) occur early and often accompany **neurological disturbances**. The latter invariably include tingling of the lips, hands and feet, **unusual temperature perception disturbances** (*cold allodynia*, a kind of burning sensation or *'pins-and-needles'*) where cold objects give a dry ice-like sensation and cause severe localized itch on the skin.

Severe cases involve hypotension with bradycardia, respiratory problems and paralysis but deaths are uncommon (less than 1%). The low fatality rate arises out of the fact that fish rarely accumulate sufficient levels of CTX to be lethal after a single meal, perhaps because fish themselves succumb to the lethal effects of higher CTX.

Diarrhea and facial rashes have been reported in **breast-fed infants** of poisoned mothers, suggesting that CTXs reach even the breast milk. **Dyspareunia** (= persistent or recurrent genital pain; just before, during or after sex) and other ciguatera symptoms have been found in healthy males and females, following sexual intercourse with partners suffering from ciguatera poisoning, also signifying that the toxin may be **sexually transmitted**.

The **phenomenon of sensitization** has also been observed, where persons, who previously were intoxicated with ciguatoxin, may suffer a recurrence of typical ciguatera symptoms after eating fish that do not cause symptoms in other persons. Such sensitization occurs many months or even years after attack of CFP. Eating fish with low levels of CTX over several years, in the absence of symptoms, eventually results in sensitization to the toxin. This is supposed to be a case of accumulation of ciguatoxin in the host or possibly an induction of an **immunological reaction**.

(d) **Treatment / Prophylaxis:**

The antidote therapy is not known. If the patient shows symptoms of ciguatera intoxication soon after ingestion of the fish, gastric lavage followed by treatment with activated charcoal helps. The biggest breakthrough in the treatment of ciguatera has been achieved with the use of '**Mannitol**' (= a diuretic, used to reduce swelling and pressure inside the eye or around the brain). It does not affect the cardiovascular or gastrointestinal symptoms but does reduce the severity and duration of neurological symptoms. Mannitol competes with sodium channels and it acts as an osmotic agent at the cellular level, thus reducing fluid excess in the cytoplasm of nerve cells or preventing influx of Na^+ through Na^+ channels to stabilise the cell membrane. Further, mannitol also reacts directly with the toxin to neutralise it or displace it from its binding site on the cell.

In the case of dehydration and hypotension, intravenous crystalloid infusion and **vasoactive agents** may be required *e.g.*, **atropine sulphate** for bradycardia and **dopamine** infusion for severe hypotension. In cases of respiratory depression, mechanical ventilation may be necessary. Calcium channel-blocker type drugs such as **Nifedipine** and **Verapamil** are effective in treating the symptoms like poor circulation and shooting pain in the chest. These symptoms are due to cramping of arterial walls caused by **MTX**. **Steroids** and **vitamin supplements** support the body's recovery rather than directly reducing toxin effects.

Apart from the avoidance of consumption of large predatory fish, the use of screening tests is the only tool presently available to prevent ciguatera intoxication. It has been suggested to avoid ingesting fish weighing more than 1.35 to 2.25 *kg*. However, there is no way of knowing the size of fish from which the steak or filet was cut. If people are educated to avoid consuming heads, viscera and roe of reef fish and told to avoid fish caught in the areas known for frequent occurrence of CTX intoxication, the incidences of ciguatera can decrease.

(x) **Haff disease or Rhabdomylosis**

[**Alternative names:** Haffkrankheit; Haff syndrome, Myoglobinuria paroxysmalis, Haff–Iuksov–Sartlan disease; Iuksovsk–Sartlansk disease]

(a) **Species/Group** (Table 1.1):

- **Osteichthyes**:
  Anguilliformes:
   Anguillidae (Freshwater eels)
  Salmoniformes:
   Salmonidae (Salmonids)
   Esocidae (Pikes)
  Cypriniformes:
   Catostomidae (Buffalo fish, Gourdhead, Redmouth Buffalo or Common Buffalo; often nicknamed as 'baldpate' because the fish has a big, bare head).
  Characiformes:
   Serrasalmidae (Pacu-manteiga, Tambaqui, Pirapitinga).
  Gadiformes:
   Gadidae (Codfishes, Burbots)
  Perciformes:
   Carangidae (Jacks, Scads, Pompanos, Yellowtail Amberjack).

(b) **Epidemiology and Toxicity:**

The said disease occurred as outbreaks after consuming **bonyfish** or **crustaceans** (shrimps), the first one being reported in 1924 in a population residing around the Northern part of the '**Vistula Lagoon**' (= '*Frisches Haff*' in German) along the the Baltic coast, hence, *Haff disease*. The recurrence of the syndrome along the Baltic coast lagoon was seasonal, mainly during summers and initially *Lota lota* (Burbot), *Anguilla anguilla* (Eel) and *Esox* sp. (Pike) were the fish found associated with this kind of toxicity. Reports of seabirds and cats dying after eating fish, are also not uncommon.

Haff disease became a deep concern for the freshwater fish consuming residents of Brazil. In an outbreak of October 2008, 27 cases of *Haff disease* were reported associated with the consumption of some big-sized **Characins** (Serrasalmidae) *viz.,* *****Pacu**-manteiga Silver Mylossoma (*Mylossoma duriventre*) [Fig. 1.8a]; **Tambaqui , Black *Pacu, Black-finned *Pacu, Giant *Pacu, Cachama** or **Gamitana** (*Colossoma macropomum*) [Fig. 1.8b] and **Pirapitinga Red-bellied *Pacu** (*Piaractus* [= *Colossoma*] *brachypomus*).

Between 2016 and 2021, a good number of cases suffering from *Haff disease* were reported from Brazil, but involving another fish, called **Yellowtail Amberjack** (*Seriola lalandi*) [Fig. 1.8 c].

> *****Pacu** = common name used to refer to omnivorous South American freshwater Serrasalmid Characins.

Alarming reports about *'Haff Disease'* came into limelight between 1984 and 1997, all after the consumption of **buffalo fish** (*Ictiobus cyprinellus*) [Fig. 1.8d], whose native and introduced distribution is confined to Canada and the United States of America. The most recent US outbreak was reported in Brooklyn (2011) and in 2019 all Illinois cases of *Haff disease* were related to consumption of buffalo fish.

The **buffalo fish** (up to 29 *kg*, 1.2 *m*) looks like a carp without barbels and has a more terminal mouth. They are usually dark coloured (brownish or blackish), often with a coppery sheen; sometimes greenish or grey, underside whitish or yellow and fins dark. It is a filter-feeder of crustacean zooplankton, sometimes feeding close to the bottom, by making up and down movements to swirl water and thus filtering plankton that hover at the bottom. They are also not averse to algae and plant parts when hungry. Sometimes this fish feeds on parts of an **aquatic weed** called **water hemlock** (*Cicuta douglasii*; Family Apiaceae), found throughout North America. The underground parts of the plant, especially the **tuberous roots**, are highly toxic.

Fig. 1.8 Haff disease-causing fishes: (a) Pacu Silver Mylossoma (*Mylossoma duriventure*); (b) Tambaqui (*Colossoma macropomum*); (c) Yellotail Amberjack (*Seriola lalandi*); (d) Buffalo Fish (*Ictiobus cyprinellus*).

THE TOXIN

Till date the origin and type of toxins involved in *Haff disease* remain unclear but it is hypothesized that **Palytoxin-like 'myotoxic'** compounds are accumulated in the body of fish and crustaceans *via* trophic transfer through the food web. Therefore, it is believed that *Haff disease* is caused by consumption of fish/crustaceans that have fed on some toxic plants. The toxic substance identified in **water hemlock** has been named as *'cicutoxin'*, an **unsaturated alcohol** having a strong carrot-like odour. It becomes more concentrated as the tubular roots of plants mature (also in the leaves and stems during early growth). This toxin is found to be very similar to **palytoxin** because the symptoms of **palytoxin-poisoning** are quite similar to *Haff disease*, both exhibiting **rhabdomyolysis**.

(c) Symptoms:

Haff disease is characterized by sudden, severe muscular stiffness, often accompanied by dark-coloured urine. Therefore, another name **'rhabdomylosis'** was given to this **'syndrome'**, owing to the disintegration of striated muscle fibers (a serious myopathy) and releasing the intracellular contents into the circulatory system.

Symptoms appear within 24 *h* of ingesting the fish/crustacean. The clinical manifestations of rhabdomyolysis are the increased **serum creatine kinase** (CK) and **myoglobin** (Mb). It can be potentially life threatening, especially if the intracellular contents are filtered by the kidneys. As the kidneys eliminate muscle metabolites, **urine** becomes reddish/ brownish/coffee coloured. The release of the **CK, Mb** and other toxic substances in turn cause injury to the kidneys and skeletal muscles *via* the bloodstream

Physical symptoms of rhabdomyolysis include profound muscular pain and weakness, swelling, stiffness and cramps. Other manifestations included bitter/metallic taste, weakness, excessive sweating, nausea, diarrhea, bradycardia, cyanosis and respiratory distress. Although deaths have been reported, most patients recover without any consequence.

(d) Treatment/Prophylaxis:

The treatment for Haff disease is primarily supportive. There is no antidote. The treatment mainly involves intravenous fluid transfusions to help the kidneys flushing out the released muscle cell contents circulating in the blood. Activated charcoal also helps, if the toxin-containing food is still in the gastrointestinal tract. Urinary alkalinization, with sodium bicarbonate infusion, has also been effective. Hemodialysis is suggested in cases with oliguria or anuria.

It is imperative that any suspected cases of Haff disease are reported to local health authorities so that other contaminated fish may be recovered and exposed persons warned of possible health risks. Recovered fish samples should also undergo for testing to identify the toxin.

1.4. GENERAL LINE OF TREATMENT/PREVENTION OF FISH POISONING:

As described with all cases of fish poisoning in the foregoing, the treatment is largely symptomatic. There are no specific antidotes and any treatment does not impart immunity. Gastric lavage and catharsis is required to be instituted at the earliest possible. In many instances, **10% Calcium Gluconate** given intravenously has resulted in prompt relief whereas in others it has been ineffective. **Paraldehyde** and **ether** inhalations have been useful in controlling convulsions. **Nikethamide** or one of the other respiratory stimulants is advisable in cases of respiratory depression. In cases where excessive mucus production takes place, aspiration is essential. Oxygen inhalation and intravenous administration of fluids supplemented with **Vitamin B complex**, given parenterally, are beneficial. In case of laryngeal spasms, intubation and tracheotomy may be necessary. When severe pain is experienced, opiates such as morphine (or any other appropriate pain-killer), given in small divided doses, is required. Cool showers have been found to be effective in relieving severe itching. Patients suffering from the paradoxical sensory disturbance are given fluids, slightly warm or at room temperature. Antihistaminic drugs are found useful in the treatment of scombroid poisoning.

Complementary and alternative therapies:

The studies have shown that certain vitamins and nutrients may protect against some food toxins while others may actually worsen the effects of toxins. Homeopathy may help treat diarrhea in children (which is sometimes caused by food poisoning) in developing countries.

Nutrition:

The following general nutritional guidelines may also be helpful in the cases of fish-food poisoning:

- **To prevent dehydration, plenty of fluids are to be given.**
- **To soothe inflamed stomach or intestine, barley or rice water is recommended.**
- **Taking probiotics, such as *Lactobacillus acidophilus* and *L. bulgaricus*, helps restoring good bacterial flora in the intestine.**

Chapter – 2
VENOMOUS FISHES
Pterois volitans, a watercolour illustration.

CHAPTER 2

VENOMOUS FISHES

HIGHLIGHTS

NUMBER OF VENOMOUS FISHES
DISTRIBUTION/HABITAT PREFERENCES
EPIDEMIOLOGY, IN GENERAL
THE DIVERSITY OF VENOMOUS FISHES

Cartilaginous fishes
Chimaeras
Sharks
Sting Rays
Bony fishes
Catfishes
Spiny-rayed bony fishes

Fishes use a myriad of defenses against predation, including **toxins** and **venoms**, some produced by the fish themselves, others by symbiotic bacteria. While few of the venoms are lethal to humans, most are painful enough to deter potential predators. The venomous fishes are those equipped with a set of special apparatus having **venom gland(s)** and solid or hollow hypodermic **teeth/fin spines/stings** for synthesizing and delivering the venom, respectively. Scientifically, this tactis has been called **ichthyoacanthotoxism** (*Gr. acanth*, prickle or spine).

Fish venoms are usually mixtures of heat-labile **high molecular weight proteins** with systemic toxic effect and **low molecular weight amines** causing inflammatory reactions. Although there is a complex balance between different components of the venom response, similarities exist between the responses to the venoms of all species of fish. The most potent effect of piscine venoms is on the cardiovascular system.

2.1 NUMBER OF VENOMOUS FISHES, A PHYLOGENETIC BASIS:

Fish with venomous spines are found amongst more than 200 **cartilaginous fishes** (particularly in rays and skates, less so sharks and chimaeras) and the **bony fishes**. Venomous **ray-finned bony fishes** are quite diverse with representatives spread across **four Orders** *viz.*, **Siluriformes** (Catfishes), **Batrachoidiformes** (Toadfishes), **Scorpaeniformes** (Scorpionfishes) and **Perciformes** (Surgeonfishes, Scats, Rabbitfishes, Sabertoothed blennies, Stargazers and Weeverfishes).

The only order containing venomous species resolved as **monophyletic** is that of **toadfishes** (Batrachoidiformes). The remaining spiny-rayed fish orders with venomous **scorpionfishes** (Scorpaeniformes) and **perchlike fishes** (Perciformes) are supposed to be **polyphyletic**. When a fish family tree is constructed by comparing DNA samples from species belonging to **Class – Actinopterygii** and **Superorders – Paracanthopterygii** (Batrachoidiformes) and **Acanthopterygii** (Scorpaeniformes and Perciformes) venomous fish are found scattered all over the tree. This suggests that venomosity evolved multiple times in fish.

Based on phylogenetic studies, venom apparatuses originated, independently, 11 times in spiny-rayed bony fishes. In combination with studies on the phylogeny of catfishes, it has been suggested that > 1,200 fish species are venomous and it is likely that 1,500–2,000 ray-finned fishes may be venomous when catfishes are examined in detail. Studies have indicated that the **most common venom apparatus** *i.e.*, **venom glands associated with fin spines**, has convergently evolved in 9 out of the 11 venomous spiny-rayed fish clades. The other envenomation structures (*i.e.*, **teeth, dorsal opercular spines, central opercular spines** and **cleithral spines**) are unreversed and uniquely derived.

2.2 DISTRIBUTION/HABITAT PREFERENCES :

Most venomous fish are **marine, bottom-dwellers**, living close to the coast, with the greatest variety occurring in tropical waters, particularly from Indo-Pacific, off eastern and southern Africa, Australia, Polynesia, the Philippines, Indonesia and southern Japan. They are often found in shallow waters and can also occasionally be seen in pools in intertidal zones *e.g.,* rays, weeverfishes, scorpionfishes and stonefishes. Lionfishes move freely through water and have distinctive warning colours. Many camouflage the surroundings and can also dig themselves into the loose substrate at the ocean floor. In **freshwater**, the only fish that prove to be fatal are the **catfishes**, in general; and the **freshwater rays** of South America (Potamotrygonidae).

2.3 EPIDEMIOLOGY, IN GENERAL :

The envenomations by venomous fishes cause at least 50,000 injuries per year with symptoms ranging from blisters to intense pain, fever and death. **Fatalities** are rare and have generally been caused by **stonefishes** (*Synanceia sp.*) and **stingrays**. In case of the latter, the reported fatalities are generally more often associated with large puncture wounds in vital organs than with the actual effect of the venom.

Victims are mostly fishermen who are stung while handling nets or removing fish from a hook. Bathers and divers also remain at risk, particularly in tropical regions, but in the Mediterranean, too. The victims are usually stung in the foot by bottom-dwelling venomous fish while wading through water. However, accidents also occur when victims needlessly touch venomous fish. On the other hand, accidents may also occur outside water, in the kitchen, during cooking.

2.4 THE DIVERSITY OF VENOMOUS FISHES :

[A] Cartilaginous fishes:

(i) Chimaeras (Latin, *chimaera* = marine monster; Chondrichthyes, Holocephali, Chimaerifirmes, Chimaeridae; **Shortnose chimaeras** or **ratfishes**):

Chimaeras are marine (Atlantic and Pacific), bathydemersal and oceanodromous. Usually found in deeper waters (300 - 500 m) in southern latitudes; but while making a summer inshore migration they may be found up to 40-100 m in northern regions. They are rather sluggish, living in small groups and feeding mainly on bottom-living invertebrates. All are characterised by long (up to 1.5 m), tapering bodies and large heads. Most notably their upper jaw is fused with the skull (hence, Holocephali) and teeth fused into three pairs of hyper-mineralised tooth plates.

Chimaera monstrosa, Hydrolagus affinis and *H. colliei* are supposed to be **highly venomous**, having a unique defence mechanism in the form of a **venomous spine** located in front of the dorsal fin (Fig. 2.1 a). It may be serrated at the posterior face and surrounded by a glandular tissue. The posterior face also has a shallow groove, containing the **poisonous gland**.

On stinging, breathing disorders, hypersalivation and cyanosis are the chief symptoms. The wounds thus created are quite painful. The complete recovery takes about 6 to 48 *h*.

(ii) Sharks (Chondrichthyes, Elasmobranchii):

There are two important examples:

- ***Heterodontus sp.*** (Heterodontiformes; Heterodontidae; **Bullhead Sharks**, **Port Jackson** or **Horn Shark**):

Marine, found in tropical to warm temperate and insular shelves and uppermost slopes (0 – 275 m, most shallower than 100 m); western Indian (Arabian Peninsula to South Africa) and Pacific (Western Pacific, from Japan to Tasmania and New Zealand); eastern Pacific from California to Galpagos Islands and Peru). They are smaller (up to 1.5 m) with a blunt head, **2 dorsal fins with spines** and an anal fin. Colour pattern with a conspicuous set of harness-like narrow dark stripes on the back, are unique to the species (Fig. 2.1b).

Heterodontus portusjacksoni, H. mexicanus H. francisci are fatal as the two dorsal fins are provided with the **venom apparatus**, consisting of a spinous process just anterior to each of the dorsal fins. These spines are anchored to the dorsal tissue at their base and are surrounded by glands. The spines of juveniles are quite sharp, but those of the adults are usually blunt. The sharp spines of small individuals are an excellent defensive device. **Envenomations** occur when anglers handle struggling fish while attempting to remove the hook.

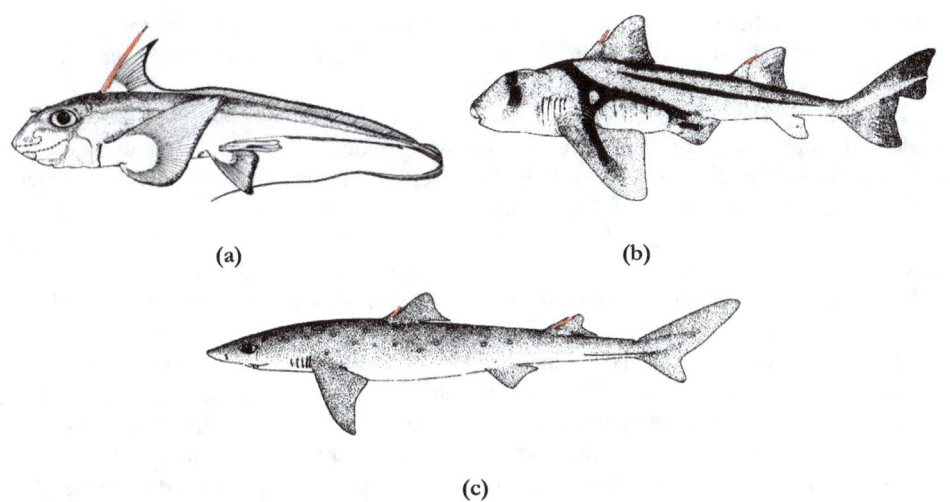

Fig. 2.1 The Cartilaginous venomous fishes: (a) Ratfish (*Chimaera monstrosa*), (b) Port Jackson Shark (*Heterodontus portusjacksoni*.), (c) Dogfish shark (*Squalus acanthias*). [**Venomous Spine** shown in 'red']

- ***Squalus* sp.** (Squaliformes; Squalidae; **Dogfish Sharks** or **Piked dogfish**):

It is marine, found in cool temperate to tropical, Atlantic, Indian and Pacific zones. Two **dorsal fins with ungrooved large spines**, white spots on grey background, oblique-cusped cutting teeth on both the jaws; absence of subterminal notch on caudal fin, anal fin and upper precaudal pit and lateral keels on caudal peduncle, are the chief diagnostics. The body is fairly slender (1.6 m). Economically, it is it is edible, yields leather, liver oil and is ground for fertilizer and fish meal. Usually, this is not dangerous in the sense of attacking people but proves hazardous to those who catch it and get inflicted due to the presence of sharp teeth and **toxic fin spines**.

As many as 34 species of Dogfish sharks have **spines** associated with the **two dorsal fins** but they appear to be associated with **toxic secretions** only in the '**spurdog**', *Squalus acanthias* (Fig. 2.1c). The poisonous glands are situated in the shallow groove on the posterior face of the spines. Envenoming is caused to fishermen after careless handling. When the spine sticks into the tissue, the poisonous glands get damaged and the poison infiltrates the wound. The dogfish can curl itself and whip its tail about to inflict wounds with the **long, sharp, second dorsal spine**. The toxin inflicts intense localised pain, swelling, numbness and muscle weakness. The hypersensitivity around the wound lasts for a few days. In some cases, the stabbing leads to death.

(iii) Sting Rays [Chondrichthyes, Elasmobranchii, Myliobatiformes; **Six-gill Stingrays** (Hexatrygonidae), **Deep water stingray** (Plesiobatidae), **Round sting rays** (Urolophidae), **American round rays** (Urotrygonidae), **Whiptail stingrays** (Dasyatidae), **River stingrays** (Potamotrygonidae), **Butterfly rays** (Gymnuridae) and **Eagle rays** (Myliobatidae)]:

Stingrays are common in coastal tropical and subtropical marine and freshwaters throughout the world; also including those found in warmer temperate oceans (*e.g., Dasyatis thetidis*) and those found in the deep ocean (*e.g., Plesiobatis daviesi*). The **river stingrays** and a number of **whiptail stingrays** (the Niger stingray) are restricted to freshwater. Most **Myliobatoids** are demersal but the stingrays and the eagle rays are pelagic. The Stingrays of Family Dasyatidae inhabit shallow water and different species have a distinctive ray shape but their colouration often makes them hard to spot unless they are swimming. While feeding they settle along the bottom, often leaving only their eyes and tail visible. All feed primarily on molluscs, crustaceans and occasionally on small fish. For marine species, coral reefs are their favourite feeding grounds.

Curiously enough, their **edibility** in not uncommon and therefore various recipes can be observed being used in many coastal areas worldwide *e.g.,* they are grilled over charcoal, then served with a spicy sauce in **Malaysia** and **Singapore**. In **Goa** (India), they are often served as part of spicy curries.

Their most sought-after edible parts are the so-called wings (pectoral fins) and the liver. Obviously, for edibility point of view their fishing becomes an important sport activity along sea coasts and so also becoming the cause of a number of **envenomations**.

The sting and the venom:

Most stingrays have **one** or **more barbed stings** (modified dermal denticles) on the tail, used exclusively for self-defense (Fig. 2.2). If broken accidentally or left in the wound, it regenerates later on. The sting is approximately 4.0 cm long whereas in larger species it may reach up to 40 cm. Encased in an integumentary sheath, it is composed of an inner core of vasodentine and the outer layer of enamel. It is bilaterally serrated, the serrations being directed cephalically.

The **venom glands** are contained in the epidermal tissue within and overlying the **two ventrolateral grooves** of the spine (Fig. 2.2 b). In **Round sting rays** (*Urolophus halleri*), the integumentary sheath is composed of inner layers of loose areolar connective tissue covered by an outer layer of epithelium. A layer of varying thickness of oval, vacuolated cells, having nucleus compressed at one end and eosinophilic cytoplasm, overlies the basal epithelium. These cells occupy 20 to 60% of the total tissue area of the **epidermis** in and over the ventrolateral grooves.

Earlier, the venom was found to be composed of **phosphodiesterase** (degrader of the phosphodiester bonds in cAMP and cGMP like second messengers), **nucleotidase** (hydrolysing nucleotides into a nucleoside and a phosphate) and **serotonin** or **5-hydroxytryptamine** (a monoamine neurotransmitter).

Recent investigations on various stingray venoms have revealed that the venom contains a wide array of components like **acetylcholine**; **proteolytic enzymes** against casein, gelatin, hyaluronic acid and fibrinogen; and the **toxins** like **cystatins** (potent inhibitors of cysteine proteases), **galectin** (having anticoagulant, myotoxic and haemagglutination properties), **peroxiredoxin** (antioxidant enzyme, controlling cytokine-induced peroxidase levels; thereby mediating signal transduction), **orpotrin** (vasoconstrictive nonapeptide, causing arteriolar constriction) and **porflan** (involved in inflammatory processes) and other peptids and proteins. **Hyaluronidase** is a latest addition, having antigenic properties and causing hydrolysis of hyaluronic acid of extracellular matrix. More importantly, **galectin** induces cell death in victims and **cystatins** inhibit defense enzymes, usually leading to increased blood flow in the superficial capillaries and cell death.

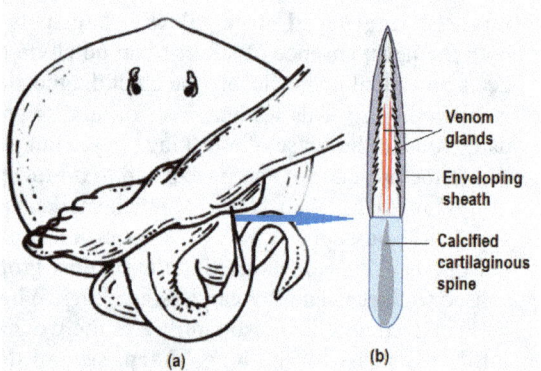

Fig. 2.2 Venomous tail spine of a Sting Ray:
(a) Stinging operation, **(b)** Isolated spine showing 2 venom glands on the ventral side.

The **toxicity of the venom** of freshwater stingrays is far greater than that of marine stingrays, because there are more proteins in rounded cells with a large amount of granule-filled cytoplasm. This difference in toxicity is also due to the fact that, the venom-producing cells are located within the lateral grooves of the 'sting' in marine stingrays whereas they branch out beyond the grooves to cover a larger area on the blade of the 'sting' in freshwater stingrays.

Envenomation:

Usually, stingrays pose a risk to people wading through water and often get injured on the leg (Fig. 2.2 a). The fishermen and divers sometimes get lashed by a startled stingray as they swim above it. Stingrays defend themselves by lashing out whip-like tails, capable of penetrating spines through wet-suits or shoe leather, thus causing serious injuries. The most common stie of injury is the lower extremity, followed by the upper ones. There are about 750 – 2000 stngray injuries reported annually from Unites States. A good number of cases are also reported from freshwater stingrays, 80-90% of injuries being to males.

Symptoms:

Bluish or greyish discolouration around the wound, disproportionate pain, swelling, muscle cramp, weakness, seizure, hypotension, cardiovascular toxicity, deep wounds and lacerations are the chief clinical symptoms. Pain normally lasts up to 48 h, but is most severe in the first 30 to 60 minutes and may be accompanied by nausea, fatigue, headache, fever and chills. There is always a possiblility of infection from bacteria at the wound site. Immediate injuries to humans include poisoning, punctures, severed arteries and death. Most of the stingray injuries have low morbidity, with the larger rates of injuries and complications in freshwater stings as compared to marine stings.

The sting ray was in news when it stung the famous 'Crocodile Hunter' Stephen Robert Irwin, the Australian naturalist and television personality, on 4th September 2006; only the second recorded in Australian waters since 1945. Steve died of heart failure due to delay in medical attention.

Treatment/prevention:

Prevention involves continuous shuffling of feet when wading. All stingray injuries should be medically assessed; the wound needs to be thoroughly cleaned with sea water and surgical exploration is often required to remove any barb fragments remaining in the wound. Hot water (43.0 – 46.0 °C) neutralizes stingray venom or may also draw out venom, which resembles jelly. It is required that water must be reheated every 10 minutes to keep it continually hot and soak the wound for 30 to 90 minutes, or as long as it takes for the pain to subside. Once relief felt, antibiotic ointment or cream should be applied at the wound and covered it with gauze.

[B] Bony fishes (Actinopterygii, the Ray-finned fishes)

(i) Catfishes (Siluriformes):

- **The number and distribution:**

As per an estimate the venomous catfish diversity is likely to equal or exceed that of all other venomous vertebrates (including other fishes). As far as distribution range is concerned, catfishes dwell inland or coastal waters of every continent except Antarctica but are particularly common and diverse in tropical South America and to a lesser degree in Africa and Asia. Based on **Siluriform phylogenies** and **toxicological assays**, approximately **1250-1625** species, belonging to at least **20 Families**, are supposed to be venomous *viz.*,

Callichthyidae: Callichthid armoured catfishes; freshwater; Panama and South America.
Amblycipitidae: Torrent catfishes; freshwater; Southern and Eastern Asia.
Akysidae: Stream catfishes; Southeastern Asia.
Pseudopimelodidae: Bumblebee catfishes; freshwater; South America.
Heptapteridae: Heptapterids; freshwater; Mexico to South America.
Cranoglanididae: Armourhead catfishes; freshwater; Asia, China and Vietnam.
Ictaluridae or **Ameiuridae:** North American catfishes; freshwater; North America.
Mochokidae: Squeakers or upside-down catfishes; freshwater; Africa.
Doradidae: Thorny catfishes; freshwater; South America.
Siluridae: Sheatfishes; freshwater; Europe and Asia.
Chacidae: Squarehead, angler or frogmouth catfishes; freshwater; Eastern India to Borneo.
Plotosidae: Eeltail catfishes; marine, brackish and freshwater; Indian ocean and western Pacific from Japan to Australia and Fiji.
Clariidae: Airbreathing catfishes; freshwater; Africa, Syria, southeastern and western Asia.
Heteropneustidae: Airsac catfishes; freshwater; Pakistan to Thailand, includinmg India Srilanka and Myanmar.
Claroteidae: Claroteids; freshwater; Africa.
Ariidae: Sea catfishes; chiefly marine but some brackish and freshwater also; worldwide, tropical to warm temperate.
Schilbidae: Schilbeid catfishes; freshwater; Africa and southern Asia.
Pangasiidae: Shark catfishes; freshwater; southern Asia, Pakistan to Borneo.
Bagridae: Bagrid catfishes; freshwater; Africa and Asia.
Pimelodidae: Long-whiskered catfishes; freshwater; Panama and South America.

- **General structural organisation of the venom apparatus:**

All catfish, belonging to above-referred families, possess a strong, hollow or solid **spiny bony ray** (= sting) on their **dorsal** and/or **pectoral fins**, covered by a thin integumentary sheath (Fig. 2.3, 2.4). These spines are locked at the site of articulation, so that they are projected outwards in a manner as to inflict severe wounds. The **venom glands** are located in a series on sharp, recurving teeth capable of cutting into a victim's flesh and facilitating the venom to be absorbed and often causing serious infections. In some fish, **unicellular glands** are the aggregates of cells, called venom glands, resembling multicellular glands of terrestrial animals. Even if covered with a connective tissue sheath, the aggregates of the venom cells do not have any common outlet.

In addition to the venom glands associated with the dorsal/pectoral spines, some catfish families are found possessing **holocrinous** secretory glands situated in the **'axils'** of the pectoral fin *e.g.,* Akysidae, Ariidae, Callichthyidae, Ictaluridae, Mochokidae, Doradidae and Plotosidae. These **axillary glands** are small, triangular, pouch-like structures, releasing their secretions through a pore located below the postcleithral process, close to the base of the pectoral fin spine.

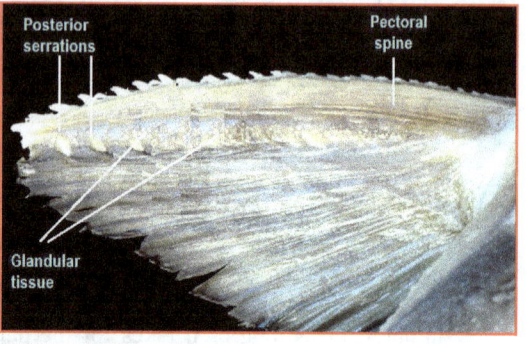

Fig. 2.3 Venom glands of a typical Doradid catfish.

Whereas most species of **Doradidae** have a single-pored **saccular axillary gland** at the pectoral spine, others possess a glandular tissue associated with several pores in the axillary region and between the posterior serrations (Fig. 2.3). *Acanthodoras* is the only genus of catfish known to secrete a milky substance through an axillary pore located just below the base of the posterior cleithral process. Microscopically, the glands are separated into discrete chambers by integumentary tissue, each chamber containing a cluster of glandular cells.

- **Catfish Venoms and Toxicity mechanisms:**

It has been hypothesized that the venom glands of fishes are derived from glandular epidermal cells that secrete toxic **proteinaceous** compounds (= **ichthyocrinotoxins**). With the exception of ichthyocrinotoxins associated with the epidermis of the dorsal and pectoral fin, there is no effective delivery device for these toxins. The venom of catfish is a complex composition of **haemolytic**, **dermonecrotic**, **oedema-producing** and **vasoconstrictive factors**, the potency of which is inversely proportional to the size of fish. **Two toxicity-causing mechanisms** have been found in catfishes *viz.,* the first being linked to **sting penetration** and rupture of the venom glandular tissue surrounding the sting whereas the second, called **crinotoxicity**, is associated with the production of toxins in the entire fish skin. Among members of the Family **Plotosidae** and in the Genus *Heteropneustes*, proteinaceous venom is so strong as a person may be hospitalized. The possible functions of the **axillary gland secretions** are supposed to be antimicrobial, ichthyotoxic, pheremonal and ionoregulatory.

- **General symptoms:**

Catfish sting envenomation is a frequent cause of morbidity among anglers, fishermen, food processors and aquarists. After envenomation there is persistent cutaneous oedema, erythema, intense burning or throbbing pain at the wound site. Paresthesias, weakness, localized sweating and muscular fibrillation can be accompanied by cyanosis and inflammation around the puncture site. Lymphangitis, cellulitis and septicaemia may also occur. Other systemic symptoms include tachycardia, hypotension, nausea and vomiting, dizziness, respiratory distress and loss of consciousness. Complications arising from secondary infection of the wound are also frequently encountered.

- **General prophylaxis/treatment:**
    i. For getting relief from pain from a sting, affected area should be immerged in water as hot as is tolerable.
    ii. Impregnated spines should be removed with tweezers.

iii. The wound should be scrubbed and washed with freshwater. The wound should not be taped or sewn together.

iv. On serious infections, oral antibiotics are usually recommended. Pain associated with a catfish sting may be relieved with one to two **acetaminophen** every four hours and/or one to two **ibuprofen** (or any latest pain-killer) every six to eight hours.

SOME IMPORTANT EXAMPLES OF VENOMOUS CATFISHES

(a) *Ictalurus* (Channel catfish) and *Noturus* (Madtoms) [Ictaluridae]:

The native range of **Channel catfish** (*Ictalurus* spp.) is the Great Lakes and St. Lawrence River watershed, the Missouri River system, the Mississippi River watershed, Gulf of Mexico watershed and parts of Mexico in North and Central America. They have also been introduced into landlocked areas of Europe and parts of Malaysia and Indonesia. Its preferred habitat is the main channels of small to large rivers, from clear, rapidly flowing, rocky-bottomed to mud-bottomed; and ponds, lakes and reservoirs.

The North American **tadpole madtom** or **stonecat** or *Noturus* (*Gk. notos* = back or drsal; *oura* = tail *i.e.,* 'back tail', meaning thereby that the adipose fin is fused along its entire length of the dorsum) is a group of small fishes (30.0 cm) in **Ictaluridae**.

In both the cases there is an inflicting **pectoral spine**. On account of longer spines, *Ictalurus* sp. (18–23 kg; 1.3 m.; Fig. 2.4 a) cause relatively deep wounds with nerve and tendon injuries. *Noturus* sp (Fig. 2.4 b) possess the **most virulent stings,** inflicting a mild to bee-like sting. **Venom glands** are present at the pectoral spine bases. Extensive tissue damage, aside from wound infections, caused by stings from these catfishes are primarily due to local effects of the venom. If the spine breaks off, parts of it may remain in the wound along with other foreign matter and in addition to causing local toxic effects and bacterial infections, can lead to poor wound healing and extensive necroses. Nausea, vomiting, diarrhoea, increased perspiration, confusion, loss of consciousness, cardiac arrhythmias, paralyses etc. are the main **symptoms**.

For **symptomatic treatment**, injection of a **local anaesthetic agent** (*e.g.,* 1% lignocaine) directly into and around the wound is a suggestive measure for pain relief. Nerve block anaesthesia with 1% **lignocaine** or preferably **bupivacaine** (a potent local anesthetic) is also recommended. **X-ray** technique is adopted for locating fragments of spine. As for adopting **first aid measures**, immersion of the affected part (up to >30 min) is required in water as hot (temp. approx. 45.0°C) as can be tolerated.

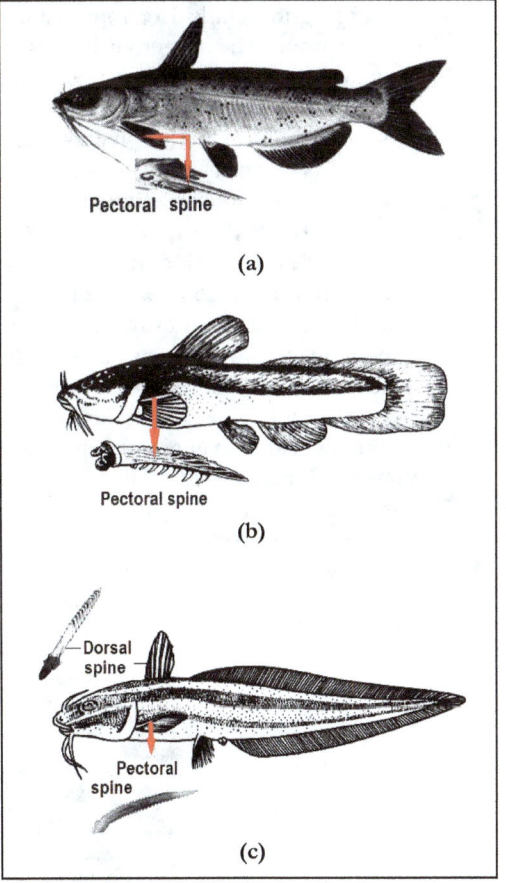

Fig. 2.4 The venomous catfishes: **(a)** The Channel catfish (*Ictalurus* sp.), **(b)** Madtom (*Noturus* sp.), **(c)** Striped Eeltail Catfish (*Plotosus lineatus*).

(b) *Plotosus lineatus* (Striped Eeltail Catfishes):

Belonging to Family **Plotosidae** (*Gk. plotos* = floating, swimming), marine and estuarine species of **eeltail catfishes** inflict stings with their venomous sharp and **serrated dorsal** and **pectoral fin**

spines (Fig. 2.4 c), covered with an integumentary sheath (like members of the Scorpaenidae). The toxicity is due to **crinotoxins** produced from their skin.

Mostly, anglers removing them from baited hooks at night, commercial fishers sorting trawl catches and unwary children handling marine animals in shallow tide pools remain at risk. After penetrating hands or feet, sometimes the spines break off leaving part of the spine in the wound, causing instant pain, redness and swelling around the wound, often turning into blue. Severe reactions include swelling of the entire limb, swelling of lymph nodes and numbness. There is always a possiblity of **gangrene,** if not treated immediately. As an immediate **first-aid**, it is a usual practice to immerse wounded part in hot water (43.0 to 45.0°C) for 30 to 40 minutes. Protective gloves should be used to remove fishes from fishing gear or trawl catches.

(c) *Clarias gariepinus* **(African catfish):**

This African catfish (Clariidae), widely introduced for aquaculture, has elongated, smooth and flexible body, with a long dorsal fin. **Stings** occur on the pectoral fins and have such articulations, as to make it erect and locked. The **venom glands** are organized as masses of **serous gland cells**, in the proximity of stings. The **epidermal secretions** are rich in phospholipids and proteins.

(d) *Heteropneustes fossilis* **(Stinging catfish)**:

Belonging to Family Heteropneustidae, *H. fossilis* (up to 70.0 cm) is the only species found in Asia. They prove to be dangerous because of the connection between the **stings in the pectoral fin** and the **large venom gland cells** lying deep in the epidermis on the sides of the spines. As in *C. gariepinus* the sting has well-developed articulations, making it possible to get erected and locked and also, the epidermal secretions are rich in lipids and proteins. **Envenomation** occurs during fishing or while handling for cooking. The sting can cause severe local pain and numbness, spreading proximally.

(ii) Spiny-rayed bony fishes:

- **Toadfish (Superorder – Paracanthopterygii, Order – Batrachoidiformes, Family Batrachoididae):**

Toadfishes (up to 25.0 cm) are marine (primarily coastal benthic; rarely entering brackishwater, a few species confined to freshwater); Atlantic, Indian and Pacific. There are **three Subfamilies** *viz.,* **Batrachiodinae** (off the coasts of America, Africa, Europe, southern Asia and Australia), **Porichthynae** (Eastern Pacific and western Atlantic) and **Thalassophryninae** (Eastern Pacific and Western Atlantic). Of them, **Thalassophryninae** (*Daector* and *Thalassophryne* sp.), are the fishes with the **highly developed venomous apparatus** of all fishes in the world. This apparatus in composed of **two hollow dorsal spines** along with a **hollow opercular spine** associated with venom glands (Fig. 2.5), capable of inflicting a painful wound to predators and unwary waders.

The chief manifestations of **evenomation** are intense local pain, erythema and edema. Some victims may develop local blisters and skin necrosis, often followed by bacterial infection. Systemic manifestations are agitation and malaise, caused by the pain. The immersion of the wounded extremity in hot water relieves the pain.

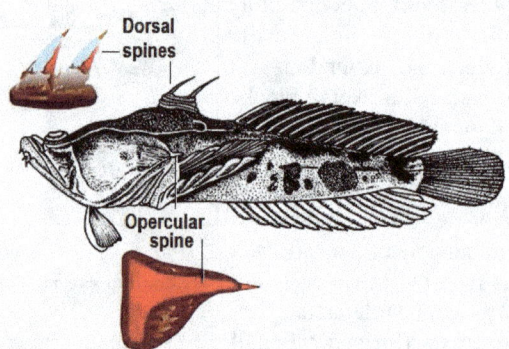

Fig. 2.5: **Toadfish** (*Thalassophryne sp.*) showing venom apparatus in the form of 2 hollow dorsal spines and 1 hollow opercular spine.

- **Scorpionfishes** or **Mail-cheeked fishes** or **rockfishes** (Superorder – Acanthopterygii, Order –Scorpaeniformes, Suborder – Scorpaenoidei, Family - Scorpaenidae):

They are brightly coloured (red, green, brown, orange, yellow, black, maroon or white but with a distinctive striped appearance and with extremely long and separated spines) belonging to at least 56 genera and 418 species.

World's most venomous species chiefly belong to **3 Subfamilies** *viz.*, **Scorpaeninae** (highly venomous **lionfishes** and **turkyfishes**: *Scorpaena, Scorpaenopsis, Sebastes, Sebasticus, Pteroi ,Parapterois, Brachypterois, Ebosia* or *Dendrochirus* etc.), **Tetraroginae** (**Sailback scorpionfishes** or **wasp fishes:** *Notesthes, Tetraroge* spp. etc) and **Synanceinae** (**Stonefishes:** *Erosa, Dampierosa, Pseudosynanceia, Synanceia* spp.).

Most species are found in coastal areas (up to about 20-60 m) of tropical to cool areas of the Indian and Pacific oceans. ***Notesthes*** is found in rivers and river mouths in eastern Australia and enter the marine environment for reproduction and from there the juvenile fish return to freshwater. All prefer rocky coasts (hence, rockfishes) and remain well-camouflaged, between stones or in crevices and amidst seaweed. Many species barely reach a length of more than 10-50 cm (*Sebastes babcocki* grows up to 64.0 cm). Head is large with spines and on each side there is a bony plate that extends from the eye up to over the gill cover (hence, the name **Mail-cheeked fishes**).

The venom apparatus:

The most common type of venom apparatus found in Scorpionfishes, is in the form of the **fin ray spines of dorsal**, **anal** and **pelvic fins**, with grooves on both sides. There are **12–18 venomous spines in the anterior dorsal fin; 1 in each of the pelvic fins** and **3 in the anal fin.**

The venom:

Proteinaceous venom granules are present in the cytoplasmic vacuoles of the gland cells *e.g.,* inflammatory **Prostaglandins** and **Thromboxane** in **Lionfishes**; **Verrucotoxin**, **Stonustoxin** (**Sto**nefish **N**ational **U**niversity of **S**ingapore; **SNTX**), **norepinephrine**, **tryptophan** and **dopamine** in **Stonefishes** and **cytolytic proteins** in **Scorpionfishes/Waspfishes**.

Evenomation:

Envenomation occurs most frequently in people accidentally stepping on the fish, the most affected ones being the fishermen, engaged in sorting out fish from the nets. When threatened, the erected spines are used for defense and the fish faces its attacker in an upside down posture to expose the spines. If the spine penetrates the flesh of the attacker or any person, the integument is pushed downwards and under the resulting pressure the venom glands discharge their contents along the grooves and into the wound. The spine may break off in victim's skin and pose a greater danger than the venom in terms of infection.

Symptoms:

Commonly, there is a severe burning pain and swelling around the wound. Less commonly caused symptoms include nausea, dizziness, muscle weakness, shortness of breath and hypotension. Rarely, blistering or tissue necrosis also occurs. Possible **bacterial sepsis** takes place if part of the spine remains in the flesh. Systemic envenomation can lead to motor paralysis, paralytic respiratory depression, hypotension and even cardiac arrest.

Treatment:

The venom is thermolabile and breaks down under heat (or perhaps heat eases out vasoconstriction). The wound, thus, be immersed in hot water (43.0 to 45.0°C) for 30 to 40 minutes. **Antivenom** is available for **stonefish** sting. It is a horse anti-stonefish toxin immunoglobulin-G with clearly established efficacy for analgesia and diminution of tissue damage due to toxin. The dose depends on the number of stings and is the same for adult and children. It is given intramuscularly and can be repeated when necessary. In case of severe systemic envenomation, it can be diluted and infused intravenously.

SOME IMPORTANT EXAMPLES OF VENOMOUS SCORPIONFISHES

(a) *Pterois volitans*:

P. volitans (5.0 to 45.0 cm) is native of the Indo-Pacific region, but has invaded the Caribbean Sea as well as East Coast of the United States. It prefers shallow waters and underwater caves. Because of its brilliant colours and long delicate fins, it can be easily located. The **venom apparatus** involves **13 dorsal, 2 pelvic** and **3 anal spines** (Fig. 2.6). On account of large, venomous spines that protrude from the body, similar to a mane of a lion, gives it the name **'lionfish'**. The venomous spines make the fish inedible or deter most potential predators. A **venom gland** is located on each side at the base of each spine (Fig. 2.6). After the spine punctures the skin, the venom enters the wound by traveling up a **shallw groove/channel** in the spine. A loose integumentary sheath covers each spine (Fig. 2.6) and during envenomation, the sheath is pushed down the spine, causing compression of two venom glands. The **neurotoxic venom** travels from the glands through grooves in the wall of the spine and into the puncture wound.

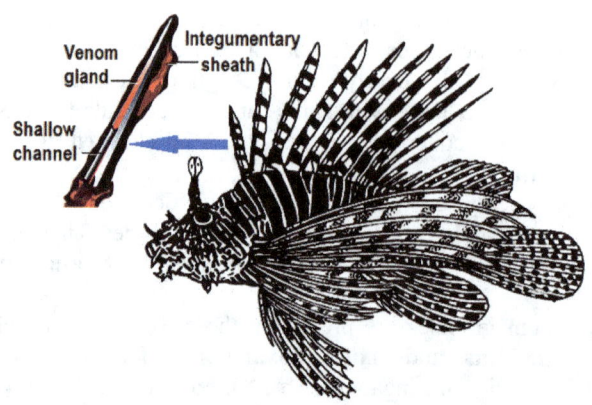

Fig. 2.6: *Pterois volitans* with venom apparatus consisting of 13 dorsal, 2 pelvic and 3 anal spines, along with enveloping integumentary sheath.

At least **four toxins** are identified *viz.,* a primary antigenic heat-labile **proteinaceous toxin**; a neurotransmitter **acetylcholine**; a **neuromuscular toxin** and a low molecular weight **non-proteinaceous ichthyotoxin**. When threatened, the fish often faces its attacker in an upside-down posture.

(b) *Synanceia horrida*:

Commonly called **Stonefish**, *Synanceia horrida* (about 30.0 cm) is found distributed around Singapore, Malaysia, Indonesia, India and as far as Africa. As with other stonefish, the scaled body has an upwardly directed head. The specifically designed pectoral fins are used to rapidly dredge sand or mud from beneath itself, allowing it to settle deeply with only its mouth and eyes fully exposed, thus making them excellent ambush predators on shrimps and small fish. Their cryptic colouration and hunting strategy makes any stonefish dangerous to any intruder or even human beings. Typically, the **venom apparatus** involves **13 dorsal, 2 pelvic** and **3 anal spines** with thick, warty integumentary sheath; **two glandular grooves in each spine** and **fusiform**

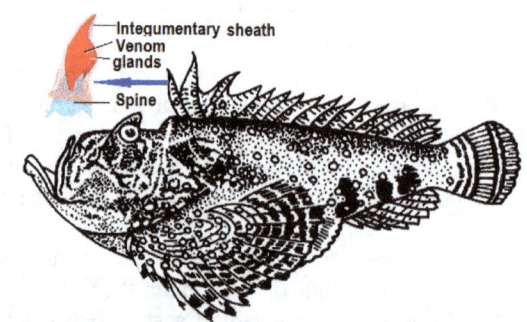

Fig. 2.7: Stonefish (*Synanceia* sp.) with venom apparatus consisting of 13 dorsal, 2 pelvic and 3 anal spines with thick, warty integumentary sheath, two glandular grooves in each spine and fusiform venom glands and their ducts within these grooves.

venom glands and their ducts within these grooves (Fig. 2.7). The dorsal spines are short, stout and straight. The distal end of each gland terminates in a duct like structure, extending from the glands to the tip of the spine.

Envenomation by stonefish usually occurs in food handlers during transfer or cooking and sting injury ranks highest in this group.

- **Weeverfishes** (Order – Perciformes, Suborder – Trachinoidei, Family – Trachinidae):

Distributed in eastern Atlantic and Black sea only two species *viz.,* monotypic ***Echiichthyes vipera*** (the lesser weever; up to 18.0 cm) and ***Trachiunus draco*** (the greater weever; up to 37.0 cm) represent the venomous **weeverfishes** or **viperfishes** (Old French *wivre* = serpent or dragon; Latin *vipera*). They are unusual in not having swim bladders and as a consequence, sink down as soon as they stop actively swimming. During the day, they lie buried in sand (Fig. 2.8), only their eyes being exposed and while doing so they cleverly attack their prey, may be a shrimp or a small fish.

The venom apparatus:

The venom apparatus consists of **1 opercular spine** and **4 to 8 dorsal spines**, all covered by glandular epithelium (Fig. 2.8). The **opercular venom apparatus** consists of two components *viz.,* the **venom gland** and the **hard component** consisting of an **opercular spine**, both encased in a membranous sheath. The **hard component** has three distinct regions *viz.,* the **opercular spine** with a **longitudinal groove** running along its length and terminating a short distance from the tip, **two basal grooves** connected with the longitudinal groove and an **opercular plate** where muscles, for raising and lowering the spine, are attached (Fig. 2.9). The **two venom glands** (Fig. 2.8) are lodged in the basal grooves of the opercular spine and short way up to the lateral groove, running along the length of the spine extending from the basal groove and terminating a short distance from the spine tip. The gland at the bottom is slightly larger than the gland at the top. The **sharp spines** laced with venom along its **dorsal fin** remain projecting out of the sand (Fig. 2.8), where it hides and inflicts agony on any unsuspecting bathers.

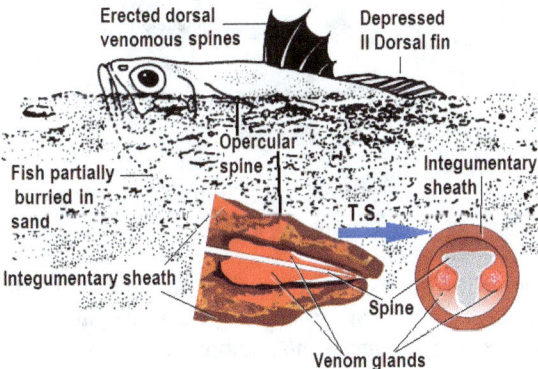

Fig. 2.8: The Weeverfish (*Trachinus* sp.): Showing the sharp dorsal spines, opercular spine and venom glands.

The Venom:

Contributing to the **venomicity** of weevers, **Trachinine** and **Dracotoxin**, respectively, are the chief venoms, provided with five major constituents *viz.,* **5-Hydroxytryptamine** (responsible for the local pain); a **histamine** (responsible for swelling), two **albumins** and a **neutral amino plysaccharaide**. The entire complex is functionally **neurotoxic**.

Evenomation:

Most human beings are inflicted by the **lesser weever**, which remains buried in sandy areas of shallow water and is thus more likely to come into contact with bathers than the **greater weever**, which prefers deeper waters. Stinging incidents are most common before and after the low tide, the frequency being increased during summers on account of the large number of bathers. Majority of injuries occur on the foot as a result of stepping on the buried fish, other common sites of injury being the hands and buttocks. Weever stings have also been known to penetrate wet suit boots even through a thin rubber sole.

While delivering the venom, the **opercular spine** is first raised to expose the spine. This is achieved by the interaction of two set of muscles (flexor and extensor) attached to the opercular plate (Fig. 2.9), working antagonistically (flexor for raising and the extensor for lowering). As the spine is raised, the outer membrane is pulled down the spine. The extent to which the tip of spine is exposed depends on the angle at which the opercular plate is raised. The outer membrane moving down the shaft builds up a pressure on the glandular mass, pressing the cells into the basal groove. The resultant pressure ruptures some of the cells forcing the venomous contents along the spinal groove.

Symptoms:

The pain from the sting is instant and often described as burning and crushing and can spread to involve the entire leg (or arm). Pain increases up to 30 minutes then subsides by 24 h, often persisting for some days. The puncture site develops redness, bruising and warmth. Infections are common due to the depth of the puncture and the unhygienic, murky, sandy or muddy water, often leading to gangrene. While symptoms of the weeverfish sting occur at the site itself, total body (also called systemic) symptoms such as numbness, fever, chills, seizures, fainting, nausea, abdominal cramps, low blood pressure, cardiac arrhythmia (irregular or extra heart beats), headache, joint aches, sweating, difficulty in breathing *etc* may also appear.

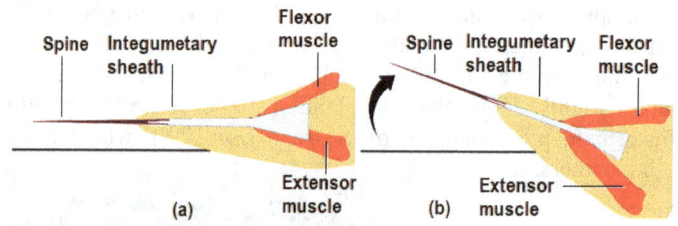

Fig. 2.9: Showing the interaction of extensor and flexor muscles for **(a)** lowering and **(b)** raising the opercular spine of a weeverfish.

Eversince 1927, when a death due to stinging by weevers was reported from UK, more stranded cases of deaths are on record *e.g.*, a teenager's death in Spain in 2020.

Treatment/prevention:

First aid treatment requires immersing the affected area in hot water, as it accelerates denaturation of the protein-based venom. Usual experience is that the pain subsides within 10 to 20 minutes, as the water cools. **Folklore** often suggests the addition of some substances to the hot water, including urine, vinegar and salt. Once the pain is eased-out, the injury should be checked for the remains of broken spines, to be removed skillfully. **Medical treatment** involves symptomic management, like use of analgesics (often with opiates). More systemic treatment using histamine antagonists assists in reducing local inflammation. Use of antibiotics is controversial, however due to the possibility of infection and the deep puncture wounds, preventative antibiotics are generally administered. Complete recovery may take a week or more and it has been reported that swelling and/or stiffness persists for months after envenomation. As precautionary measures, bathers and surfers are advised to wear sandals or wetsuit boots with relatively hard soles and avoid sitting or sporting in the shallows.

- **Stargazers** (Ordei – Perciformes, Suborder – Trachinoidei, Family – Uranoscopidae):

Uranoscopid Genera (marine; occasionally estuarine; Atlantic, Indian and Pacific) *viz.*, *Astroscopus, Genyagnus, Ichthyoscopus, Kathetostoma* (giant stargazer, 18.0 to 90.0 cm), *Pleuroscopus, Selenoscopus, Uranoscopus* and *Xenocephalus* (= *Gnathagnus*), are all demersal (at about 14.0 - 400 m) living in sandy or muddy sediments along the upper slope of the continental shelf. *Uranoscopus* spp. (Gk. *ouranos* = sky; *skopein* = to watch), having large eyes on top of the head (hence, 'stargazers'), are found worldwide, have brown-coloured and spotted dorsoventrally flattened body (up to 40.0 cm), upturned head and jaws but without a swimbladder. Both the dorsal and anal fins are relatively long; some lack dorsal spines. While burried in sand, they leap upwards to ambush prey (benthic fish and invertebrates) that happen to pass overhead. Some species have a **worm-shaped lure** growing out of the floors of the mouth which is wriggled to attract the prey (Fig. 2.10).

The venom apparatus:

The venom apparatus consists of the **cleithral** or **shoulder spines**, with venom glands and the enveloping integumentary sheaths (Fig. 2.10). The spines are sharp, conical, acellular structures of cementum-like material having concentric growth laminae. They protrude from the superior posterior edge of the operculum above the pectoral fin. The distal portion of each spine is almost rounded, while the proximal one is flattened in the medial lateral plane. There is a shallow groove along the ventral border of each spine. A similar, but much less marked groove is also found along the dorsal edge. In the integumentary sheath, the dermis is seperated from the epidermis by a basement membrane. It is relatively thick, moderately dense and has spaces filled with a gelatinous material, supposed to be the **venom**.

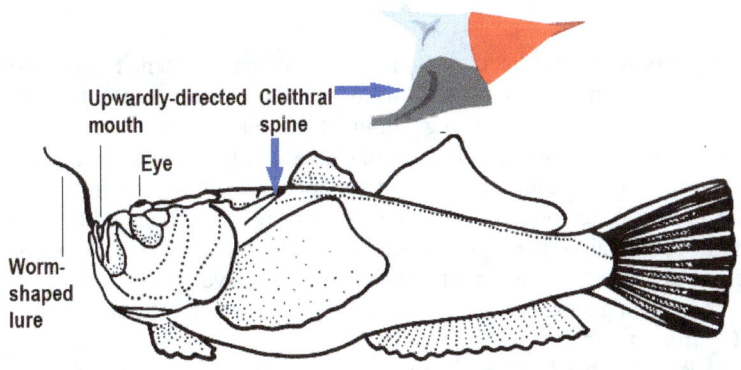

Fig. 2.9: Stargazer (*Uranoscopus* sp.): Showing cleithral spine, upwardly directed mouth, sky-facing eyes and a worm-shaped lure growing out of the floor of mouth.

Evenomation:

During stinging act when pressure is exerted upon the sheath, the venom containing cells as well as the sheath rupture and the venom is forced into the groove of the spine and then to the tip.

Symptoms:

Painful puncture wounds are accompanied by swelling, redness and heat.

Treatment / prevention:

The wound be cleaned thoroughly, bleeding be allowed for sometime and be soaked in hot water for 30-90 minutes. Often anti-tetanus treatment is recommended. The fish should be handled carefully when removing from nets or hooks.

- **Saber-toothed Blennies** (Order – Perciformes, Suborder – Blennioidei, Family – Blenniidae):

These are a specialized group of marine fishes (rarely freshwater and occasionally brackish, primarily tropical and subtropical); Atlantic, Indian and Pacific, called **comb-toothed blennies**, represented by genera like *Aspidontus, Meiacanthus, Petroscirtes, Plagiotremus* and *Xiphasia* etc. Of them, *Meiacanthus* spp. (*Gk. meion* = less = lessen, *akantha* = thorn; referring to most species having relatively few dorsal fin spines), the **sabretooth blennies** or **fang blennies**, are unique in having **toxic buccal glands** associated with **canine-like teeth on the dentary** (Fig. 2.11). Some fishes that mimic *Meiacanthus anema* attest to the effectiveness of the **toxic buccal glands** in providing immunity from predation. They inhabit sheltered shallow regions amongst reefs. On account of distinctive colouration, accentuated by their sleek body (up to 11.0 cm), they are famous in aquarium trade.

The two enlarged, **grooved canine teeth,** for which '**fang blennies**' get their name, are placed on the dentary of the lower jaw (Fig. 2.11). Unlike other fang blenny genera, *Meiacanthus* employs its

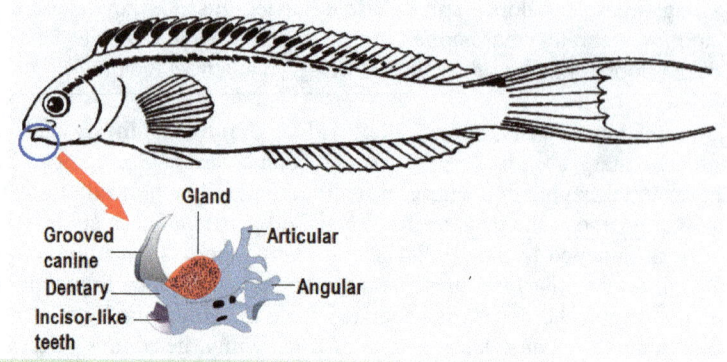

Fig. 2.11: Sabretooth blenny (*Meiacanthus anema*): Showing grooved canine on the dentary with toxic gland at the base.

weaponry to defend itself or its territory. When threatened, jaws are wide opened to expose its teeth. The holocrinous **venomous buccal glands**, situated at the base of the grooved **canines** (Fig. 2.11), are formed due to an invagination of the buccal epithelium. Slightest pressure on the glands exudes a yellowish secretion, to be channeled along the groove of the canine. In contrast to other fish bites, the blenny's venom does not cause pain, instead the venom acts like **heroin** or other **opioids** (opioid peptides), temporarily stunning or slowing the actions of the predator (as also found in the venom of cone snails). Fortunately, because of small mouth, envenomation to humans is unlikely. But if there is any, the bite causes a localised pain. However, some people may be allergic to the venom. While maintaining an aquarium, one should never attempt to capture or feed a fang blenny.

Treatment is symptomatic.

- **Rabbitfishes** (Order – Perciformes, Suborder – Acanthuroidei, Family – Siganidae):

Belonging to the only Genus – ***Siganus*** (about 29 spp.), **Rabbitfishes** (on account of their rabbit-like snout) or **spinefoots** (on account of the spines on the pelvic fins) are marine (rarely estuarine), tropical Indo-West Pacific and Eastern Mediterranean fishes. All are diurnal herbivores, some living in schools while others being solitary among the corals. Body (max. 50.0 cm) is compressed laterally and brightly coloured with complex of spots and maze-like patterns (Fig. 2.12).

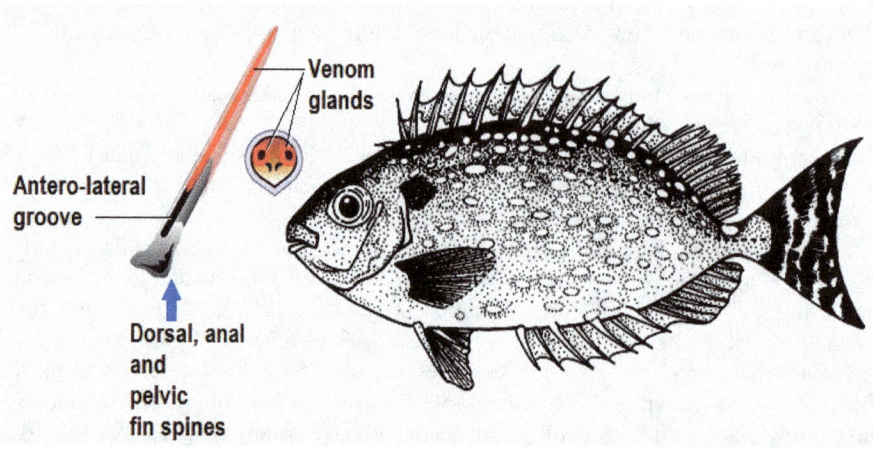

Fig. 2.12: Rabbitfish (*Siganus sp.*): Showing venom glands found in grooves within the distal 1/3rd of the fin ray spines of dorsal, pelvic and anal fins.

For the market, they are ofen used to prepare *bagoong* (a Philippine condiment made of partially or completely fermented fish or shrimps and salt). An unusual feature among rabbitfishes is their pelvic fins having 2 spines, with 3 soft rays between them. The dorsal fin has 13 spines with 10 soft rays behind, while the anal fin has 7 spines and 9 rays behind. All fin spines are grooved.

The **venom glands** are found in grooves within the **distal 1/3rd of the fin rays of all the fins** (Fig. 2.12). The spines are covered with an integumentary layer that is ruptured after trauma, releasing the venom. Their venom is not life-threatening to human beings. Most injuries occur when people handle the fish without wearing gloves. Care must be taken while maintaining an aquarium, as they are easily frightened and use their venomous spines in defense.Wounds inflicted, introduce venom into the surrounding tissues and are extremely painful. The severity of the sting depends on the number of wounds and the amount of venom injected. In severe cases, affected limbs become extremely swollen. Occasionally the pain causes delirium. Anyone, seriously envenomated, is advised to seek immediate medical assistance, until pain is under control and the condition is stabilised.

Treatment is symptomatic.

- **Surgeonfishes** (Order – Perciformes, Suborder – Acanthuroidei, Family –Acanthuridae):

They all are marine, found in tropical and subtropical seas and are quite popular as aquarium fish on account of their brilliant colour patterns. The presence of **scalpel sharp caudal blades** or **keeled peduncular plates** (Fig. 2.13 & 2.14) is diagnostic to these fishes and is the basis of calling them **'surgeonfishes'**. These **'blades'**, may be **venomous**, varying in size and number from species to species and are capable of inflicting painful wounds/lacerations. Laterally compressed and disc-like, yellow, black, dark blue or white-bodied Surgeonfishes are algae-eaters, cleaning the coral reef with their teeth.

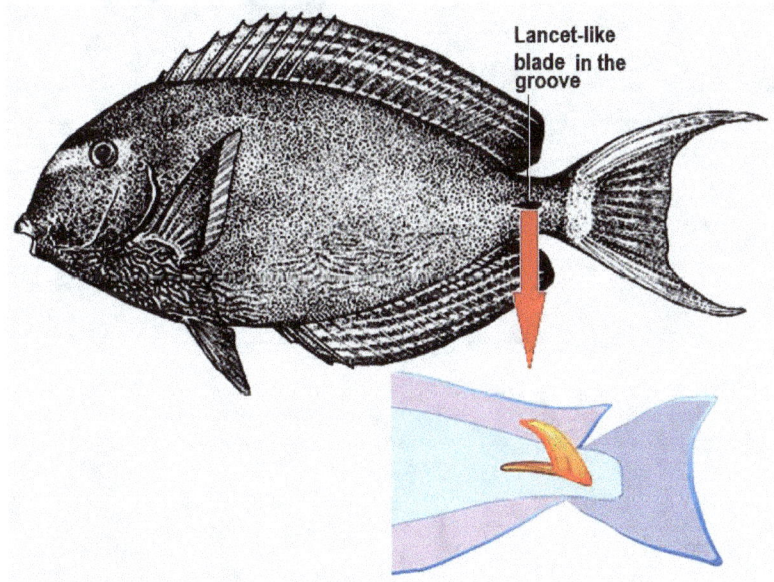

Fig. 2.13: Surgeofish (*Acanthurus sp.*): Showing erectile razor-sharp, lancet-like spine at the base of the tail.

Two subfamilies *viz.*, **Acanthurinae** and **Nasinae** are, respectively, distinguished on the basis of **one lancet-like** (Fig. 2.13) and **2 fixed keel-like spines** (may be orange or blue) (Fig. 2.14). The former includes as many as **6 genera** *viz., Prionurus, Paracanthus, Zebrasoma, Acanthurus* and *Ctenochaetus*, whereas the latter is represented by only **1 Genus** (*Naso*) and 16 species, of which *Naso unicornis* is unique in having a **'rostral protuberance'**, a hornlike extension of the forehead (hence, **'*unicornis*'**).

When the fish is excited, the **caudal spines** of **Acanthurinae** are protruded from a groove. As these species swim calmly, the blade is folded inside the groove but the moment they are threatened it protrudes out, pointing forward and ready to inflict a serious cut. In this act, they swim beside the intruder, swinging their tails to inflict long, deep-slicing cuts. When attempting to remove the fish from a net or handling it in the aquarium, deep wounds on the hands are created, resulting in swelling and discolouration, sometimes posing a risk of infection.

In *Naso unicornis* these **razor like blades** are immovable and are quite venomous (Fig. 2.14).

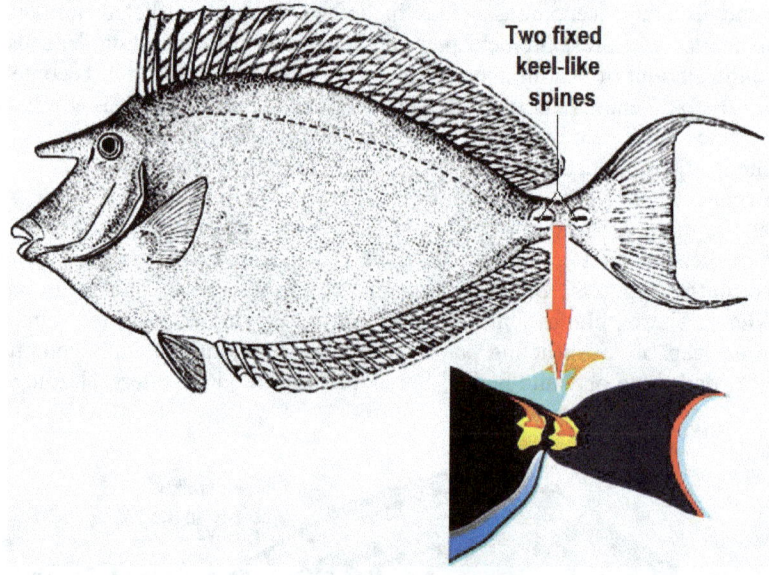

Fig. 2.14: **Surgeofish** *(Naso sp.)*: Showing 2 fixed, keel-like spines at the base of the tail.

Chapter – 3
POISONOUS AND VENOMOUS SHELLFISHES (MOLLUSCA)

Greater blue-ringed octopus: *Hapalochlaena fasciata,* a watercolour illustration.

CHAPTER 3
POISONOUS AND VENOMOUS SHELLFISHES
[Mollusca]

HIGHLIGHTS
BIVALVE SHELLFISH-POISONING
Paralytic Shellfish Poisoning (PSP)
Diarrhoeic Shellfish Poisoning (DSP)
Amnesic Shellfish Poisoning (ASP)
Neurotoxic Shellfish Poisoning (NSP)
Azaspiracid Shellfish Poisoning (AZP)
Ciguatera Shellfish Poisoning (CSP)

GASTROPOD SHELLFISH-POISONING
Ttetrodotoxin-Poisoning
Ciguatera Poisoning
Paralytic Shellfish Poisoning
Diarrhoeic Shellfish Poisoning
Neurotoxic Shellfish Poisoning
Tetramine Poisoning
CEPHALOPOD SHELLFISH-POISONING
VENOMOUS MOLLUSCS

Molluscs are important in a variety of ways *viz.*, as food, for decoration, in jewelry and in scientific studies. Edible species of molluscs include various **Bivalves** (Clams, Mussels, Oysters, Cockles, Scallops etc.), **Gastropods** (marine and land snails) and **Cephalopods** (Squid, Cuttlefish and Octopuses). Most often, eating molluscs is associated with a risk of **food poisoning** from **toxins** that accumulate in molluscs under certain conditions.

Besides, some members of **Gastropods** and **Cephalopods** are infamous of being **'venomous'**, too.

3.1 BIVALVE SHELLFISH-POISONING:

Bivalve molluscs (Clams, Mussels, Oysters, Cockles and Scallops) [PLATE IV] are **filter-feeders** *i.e.*, feeding by filtering suspended particles and phytoplankton. **Bivalve-poisonings** occur when algal/phytoplankton toxins (= **Phycotoxins** produced by some dinoflagellates and diatoms) are transferred through the food chain to higher trophic levels, including human beings. Among scientific community, mass proliferation of toxic phytoplankton is popular as **Harmful Algal Bloom(s)** or **HABs**. The **Harmful Algal Event Database** (*HAEDAT*; http://haedat.iode.org) is the only existing open access meta database holding information about harmful algal events across the world **(for details pl. refer to APPENDIX – I)**. '**Cultural eutrophication**' from domestic, industrial and agricultural wastes is the chief contributor to the **HABs**, producing noxious **toxins**.

Most common poisoning types, associated with exposure to **HAB toxins**, include the following (Table 3.1):

- **Paralytic Shellfish Poisoning (PSP).**
- **Diarrhoeic Shellfish Poisoning (DSP).**
- **Amnesic Shellfish Poisoning (ASP).**
- **Neurotoxic Shellfish Poisoning (NSP).**

- **Azaspiracid Shellfish Poisoning (AZP).**
- Ciguatera Shellfish Poisoning (CSP).

3.1.1 Paralytic Shellfish Poisoning (PSP):

(i) The Shellfish:
Bivalve molluscs, largely including mussels, clams, giant calms, oysters, scallops and cockles (PLATE IV) are identified involved in PSP.

Table 3.1: The shellfish-poisoning and the marine toxins derived from harmful algal blooms

TYPE OF POISONING	HARMFUL BLOOM-FORMING DINOFLAGELLATE(S)/DIATOM(S) [PLATE III, Fig. 3.1]	THE TOXIN	SHELLFISHES [PLATE IV, V; Fig. 3.1, 3.2]
Paralytic Shellfish Poisoning (PSP)	*Alexandrium* sp., *Gymnodinium* sp, *Pyrodinium* sp,	Saxitoxins (STXs)	Clams, mussels, oysters, cockles, gastropods (whelks), scallops, Cephalopods, lobsters, copepods, crabs
Diarrhoeic Shellfish Poisoning (DSP)	*Dinophysis* sp., *Prorocentrum* sp., *Protoceratium* sp., *Coolia* sp., *Protoperidium* sp.	Okadaic acid, Dinophysis toxins (DTXs), Yessotoxins (YTXs) and Pectenotoxins (PTXs)	Mussels, scallops, clams, gastropods,
Amnesic Shellfish Poisoning (ASP)	*Pseudo-nitzschia multiseries*	Domoic acid	Mussels, clams, Cephalopods
Neurotoxic Shellfish Poisoning (NSP)	*Gymnodinium breve* (=*Karenia brevis*)	Brevetoxins	Oyster, clams, mussels, cockles, whelks
Azaspiracid Shellfish Poisoning (AZP)	*Protoperidinium crassipes*	Azaspiracids (AZAs)	Mussels, oyster
Ciguatera Shellfish Poisoning (CSP)	*Gambierdiscus*	Ciguatoxin (CTX)	Giant Clam (*Tridacna maxima*)

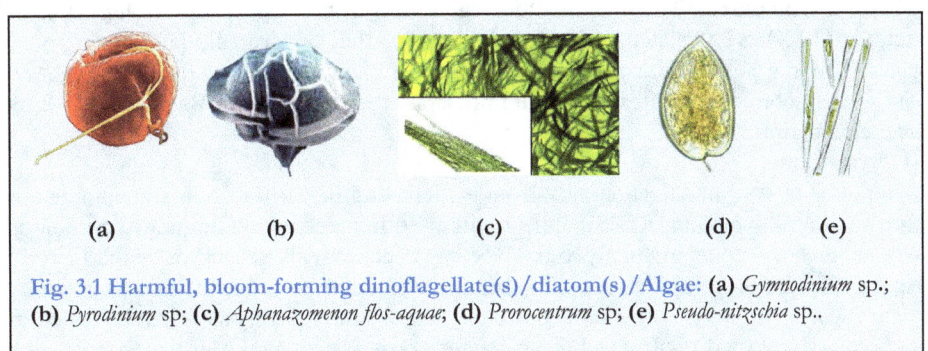

Fig. 3.1 Harmful, bloom-forming dinoflagellate(s)/diatom(s)/Algae: **(a)** *Gymnodinium* sp.; **(b)** *Pyrodinium* sp; **(c)** *Aphanazomenon flos-aquae*; **(d)** *Prorocentrum* sp; **(e)** *Pseudo-nitzschia* sp..

(ii) Epidemiology:
PSP is the most common and **most severe** of the shellfish intoxications, reported in northeastern Canada more than 100 years ago. In United States PSP has been prevalent along the coasts of New

England, the Pacific Northwest and Alaska, particularly in New England, where the first documented **PSP** case dates back to 1958. By 1990, **PSP** was well-documented from throughout the Southern Hemisphere in South Africa, Australia, India, Thailand, Brunei Darussalam, Sabah (Malaysia), the Philippines and Papua New Guinea.

(iii) The Toxins and their source:

The **PSP** toxins are identified in some **dinoflagellates** *e.g., Alexandrium* (PLATE III), *Gymnodinium* and *Pyrodinium* (Fig. 3.1 a,b) are identified as contaminators of shellfish through a toxin called '**saxitoxin**' (STX). The immobile forms of dinoflagellates, the **resting cyst** or the **hypnozygote**, are supposed to be highly toxic. Until 1970, PSP-producing dinoflagellate blooms of *Alexandrium tamarensis* and *A. catenella* were only known from temperate waters of Europe, North America and Japan. However, outbreaks of **PSP** in Japan, the northwest coast of North America, southern Ireland, Spain, Mexico, Argentina and Tasmania (Australia) have been caused by blooms of *Gymnodinium catenatum*.

A **freshwater Cyanobacteria** (*Aphanazomenon flos-aquae*) [Fig. 3.1 c] has also been shown to contain **saxitoxins** (STX) and other most toxic substance named **neosaxitoxin** (neoSTX).

> **BOX 3.1**
> **OTHER ORGANISMS CONTAINING PSP TOXINS**
>
> Copepods (*Acartia tonsa, Eurytemora herdmani* etc), Xanthid crabs (*Lophozozymus pictor*), Horseshoe crabs (pl. see ahead), Star fish (*Asterias amurensis*) also contain **PSP** toxins. Atlantic mackerel (*Scomber scombrus*) are lethal vectors of PSP toxins to predators. Besides Tetraodotoxin (TTX), marine puffers (*Arothron* spp.) from Philippines have been found to contain considerable amounts of **PSP** toxins (majorly **STX**). Freshwater puffers (*Tetraodon leiurus complex, Tetraodon suvatii*), collected from the northeastern province of Thailand are also found to contain PSP-toxins.

The term *'<u>saxitoxin</u>' originates from the specific name of the **Alaska butterclam** *i.e., Saxidomus giganteus* (PLATE IV c). S**axitoxin** actually refers to all the related **neurotoxins** produced by the dinoflagellates/cynophytes, which include pure **saxitoxin** (STX), **neosaxitoxin** (neoSTX), **gonyautoxins** (GTX) and **decarbamoylsaxitoxin** (dcSTX). At least 21 **PSP toxins** are closely related to *3,4,6 - trialkyltetrahydropurine* compounds. Most of them are heat stable at acidic pH but unstable and easily oxidized under alkaline conditions.

During filter-feeding, the dinoflagellate cells and cysts are transported to the oesophagus and the stomach of the bivalves. **PSP** toxins are released after digestion and retained in the viscera. Digestive gland, mantle, gonad and gill tissues all retain the toxins, their levels varying from tissue to tissue and species to species. There are seasonal variations, too, in toxicity level of the various tissues. After uptake and distribution, the toxins may undergo further transformation.

* <u>Saxitoxin</u> is listed as a grade one chemical weapon under the UN Chemical Weapons Convention and was used by CIA in suicide pills.

(iv) Effect of toxins:

STX is a **neurotoxin**. The **pharmacological action** of the **PSP** toxins strongly resembles that of tetrodotoxin (TTX). **STX** related toxins act *via* blocking the Na^+ channel, thus slowing or stopping the propagation of the action potential. However, the K^+ channel remains unaffected. Since STXs are charged, water-soluble molecules and do not penetrate the **Blood Brain Barrier** (BBB), most of their effects are on peripheral nerves.

(v) Symptoms:

Symptoms of **PSP** intoxication vary from a slight **tingling** or **numbness** to complete **respiratory paralysis**. Clinical symptoms of PSP appear within 30 minutes with a tingling sensation or numbness around lips, then spreading gradually to the face and neck. **Prickly sensation** on the fingertips and toes is frequent. Headache, dizziness, nausea, vomiting and diarrhea and sometimes, temporary blindness are also caused. In severe cases of poisoning, **muscular paralysis** spreads faster and deeper. Pronounced respiratory difficulty and death through **respiratory paralysis** may occur within 2 to 24 *h* after ingestion of about 1.0 – 4.0 *mg* STX, however, the **mortality rate** by **PSP** varies considerably. In recent outbreaks involving over 200 people in North America and Western Europe, no death occurred. However, in similar outbreaks in Southeast Asia and Latin America, death rates of 2 to 14% have been recorded.

Algal PSP toxins also cause **mortalities among fish** through involvement in the marine food web as evidenced by tons of **herrings** dying in the Bay of Fundy after consuming small planktonic snails that had been feeding on *Alexandrium*. From the human health point of view, it is fortunate that herring, cod, salmon and other commercial fish are sensitive to PSP toxins and, unlike shellfish, die before the toxins reach dangerous levels in their flesh. Some toxins, however, accumulate in the liver and other organs of the fish and therefore, other fish, marine mammals and birds that consume whole fish, including the viscera, are at risk. In 1987, 14 **humpback whales** died suddenly from exposure to a bloom of *Alexandrium tamarensis* in Cape Cod Bay (Massachusetts).

(vi) Treatment:

If the patient survives 18 h, the prognosis is good, with chances of complete and rapid recovery. Even nine hours appeared adequate for a physiological reduction of the toxins concentration to relatively harmless levels, except in those cases where the toxin concentration began at exceptionally high levels or in victims with impaired renal function. **Fluid therapy** is essential to correct any possible acidosis and renal excretion of the toxins.

Artificial ventilation and **gastric lavages** are still the only acceptable medical remedies against STX intoxication. However, in cases of severe intoxication, artificial ventilation may be inadequate. If no vomiting had occurred spontaneously, induced emesis or gastric lavage is used to remove unabsorbed toxins. As the **PSP** toxins are strongly charged at the gastric pH, they would be effectively absorbed by **activated charcoal**. As ventilatory failure is due to varying degrees of paralysis of the respiratory nerves and muscles, **positive pressure assisted ventilation** is desirable. *4-aminopyridine* ($C_5H_4N-NH_2$), a non-selective K^+ channel blocker that binds on the cytoplasmic side of the cell membrane, is found useful as **antidote** for STX-intoxication.

(vii) Prophylaxis:

- **Detoxification (= depuration):**

Detoxifying shellfish, contaminated with PSP, is one of the most practiced prophylaxis. In the most popular method, shellfish are transfered to waters free of toxic organisms and allow them to self-depurate. However, transferring large quantities of shellfish is labour intensive and costly. The rate of detoxification varies considerably and some species remain toxic for extended periods. Through combined efforts of an intensive monitoring programme and culture of '**rapid release**' species (*e.g.,* Blue Mussel, *Mytilus edulis*) [PLATE IVa] and the species known to avoid toxic dinoflagellates (*e.g.,* northern quahog or hard clam *Mercenaria* sp.), economic loss can be kept to a minimum.

Cooking has been proposed as a possible means of detoxifying shellfish contaminated with PSP. However, while cooking may reduce toxin levels, it does not eliminate the danger of intoxication. If the initial level of toxicity is low, cooking may effectively reduce toxicity to safe levels. **Pan frying** seems to be more effective than other methods of cooking. When clams or mussels are steamed or boiled, toxins lost from the tissues are left in the cooking liquid, rendering the fluids extremely toxic. **Boiling** of oysters (at 98°C for 10 minutes), reduces their toxicity by 68 to 81%. However, boiling by itself is not sufficient to detoxify highly toxic shellfish.

Prevention can be strengthened not only through a PSP monitoring programme but also through a strategy of **specific training to specific target groups** and an appropriate information dissemination system.

- **Regulations:**

Several countries have established or proposed **regulations** for **PSP**. Most regulations are set for PSP toxins as a group. In general, limits have often been set at 400 MU/100 g or 80 μg STX eq/100 g.

3.1.2 Diarrhoeic (or Diarrhetic) Shellfish Poisoning (DSP):

(i) The Shellfish:

Like PSP, **Diarrhoeic Shellfish Poisoning** (DSP) is also caused by ingesting contaminated **mussels, scallops, oysters** or **clams** (PLATE IV a-h), relatively highest toxicity being in **blue mussels** (*Mytilus edulis*).

(ii) Epidemiology:

The first reported cases of **DSP** were in Netherlands during 1960s, followed by similar reports in the late 1970s from Japan. In Spain, over 5000 cases were reported in 1981; in France over 2000 cases during 1984 and 1986 and over 300 cases in Scandinavia in 1984. Mussels exported from Denmark to France caused DSP poisoning in over 400 people in 1990. A two year study (1984-1986) in **India** showed that DSP toxins were present in several shellfish examined. A report of a large outbreak of **DSP** in 2000, involving 120 people in northern Greece, who consumed mussels harvested from the

Examples of bivalve shellfish-poisoning: **(a)** Blue Mussel (*Mytilus edulis*); **(b)** Green shell mussel (*Perna canaliculus*); **(c)** Alaska butterclam (*Saxidomus giganteus*); **(d)** Soft-shell clam (*Mya arenaria*); **(e)** Razor clam (*Siliqua patula*); **(f)** Giant clam (*Tridacna maxima*); **(g)** Oyster (*Crassostrea* sp.); **(h)** Scallop (*Pecten maximus*); **(i)** Cockle (*Cardium* sp.).

Adriatic Sea following algal blooms, illustrates the international concern about the shellfish poisoning outbreaks.

(iii) The toxins and their source:

Isolated from various shellfish and dinoflagellates, the **DSP** toxins are heat-stable **polyether** and **lipophilic** compounds. Depending on chemical structure, **DSP** toxins are often divided into three groups. The **first group** of acidic toxins includes **Okadaic Acid** (OA) and its derivatives named as **dynophysis toxins** (DTXs) from a marine dinoflagellate (*Dinophysis* spp.) [PLATE III d]. **OA** was named after a marine sponge *Halichondria okadai*, from which it was isolated for the first time. The **second group** of neutral toxins consists of **polyether-lactones** of the **Pecteno Toxin group** (PTXs). The **third group** includes a **sulphated polyether** and its derivatives, the **Yessotoxins** (YTXs).

A case of human intoxication, with DSP symptoms, following the consumption of mussels from Killary, Ireland in 1995 was resolved by the isolation of a **new toxin**, tentatively named **Killary Toxin-3** (KT_3). This toxin was later called **Azaspiracid** (pl. see AZP ahead).

DSP toxins are primarily produced by dinoflagellate blooms of *Dinophysis* spp., and *Prorocentrum* spp. (Fig. 3.1 d) in early spring. Toxicity of *Dinophysis* sp. varies spatially and temporally and the number of cells/litre needed to contaminate a shellfish is highly variable *e.g., D. fortii*, @ 200 cells/*l*, in mussels and scallops becomes toxic to human beings. **OA** is found in predatory fish as a consequence of their preying on mussels and fish contaminated with **OA**.

The cage-cultivated Cods, fed on toxic mussels, showed the highest concentrations of **OA**, particularly in liver while lower concentrations were found in muscle and gonads. Although most of the total toxin burden has been found associated with fatty visceral tissue, high toxin levels are observed in gonads of adult scallops.

(iv) Effect of toxins and Symptoms:

Symptoms appear 30 *minutes* to 6 *h* after ingestion of contaminated shellfish, lasting for up to 4 days. They mainly include diarrhoea, abdominal cramps, nausea, vomiting, weakness and chills. **Diarrhoea** is caused by hyper-phosphorylation of proteins that control **sodium secretion** by intestinal cells or by increased phosphorylation of cytoskeletal or junctional moieties that regulate solute permeability, resulting in passive loss of fluids.

OA also triggers **long-lasting contraction of smooth muscles** of arteries due to inhibition of myosin light chain phosphatase. **OA** is a proved to be potent inhibitor of the serine/threonine phosphatases, a critical group of enzymes closely-linked with many crucial metabolic processes within a cell. There are also concerns that **OA** may be neurotoxic, immunotoxic and embryotoxic. In addition, OA is found to be a **tumor promoter**, thus increasing risks of various forms of **cancer**.

Shellfish containing more than 2.0 *mg* **OA**/g and/or more than 1.8 *mg* **DTX**/g of hepatopancreas are considered unfit for human consumption.

(v) Treatment:

Treatment is symptomatic and supportive with reference to short-term diarrhoea and accompanying fluid and electrolyte losses. In general, hospitalization is not necessary; fluid and electrolytes can be administered orally.

(vii) Prophylaxis:

- **Detoxification:**

The rate of **removal of DSP toxin** from shellfish depends on the species concerned and may be affected by factors like feeding, temperature, salinity and the density of non-toxic algae. Low water temperatures slow down the toxin loss. It is highly dependent on the site of toxin storage *i.e.*, toxins in the gastrointestinal tract of *Mytilus* spp. are eliminated more readily than toxins bound to tissues. All methods tested until now (pl. refer to PSP in the foregoing) appeared unsafe, too slow, economically unfeasible or yielded products unacceptable in appearance and taste. Only after **rigorous boiling** (at 100°C for about 2.5 *h*), denaturation of the toxin is observed.

- **Prevention:**

Frequent inspection of seawater around aquaculture facilities or shellfish farms for the presence of toxin producing strains of phytoplankton is a right approach. Data on the occurrence, type and concentrations of toxic algal species will indicate which toxins may be expected during periods of algal blooms and which seafood products should be considered for analytical monitoring. In some countries, the shellfish cultivation areas are closed when the number of cells of certain algal species exceeds safe concentrations. Some countries close their harvest areas only when the toxins have been detected in the shellfish.The principal strategy to prevent DSP intoxication is effective monitoring of mussels with respect to DSP toxins so that contaminated products do not reach market.

- **Regulations:**

Various countries have laid down rules and regulations to have allowable levels of toxins. European Commission has directed that maximum level of OA, DTXs and PTXs together, in edible tissues of molluscs, echinoderms, tunicates and marine gastropods should be 160 *mg* **OA** equivalents/*kg*. In USA the area of domestic food safety, cooperative programmes between the **Food and Drug Administration** (FDA) and individual states exist; the FDA laying down 0.2 *ppm* **OA** as an action level. The **National Shellfish Sanitation Programme** (NSSP) provides guidelines for these cooperative agreements.

3.1.3 Amnesic Shellfish Poisoning (ASP) or Domoic Acid Poisoning (DAP):

(i) The Shellfish:

Blue mussels (*Mytilus edulis*), **soft-shell clams** (*Mya arenaria*), **razor clams** (*Siliqua patula*) and **scallops** (*Pecten maximus*) [PLATE IV a,d,e,h] have been found associated with ASP due to **Domoic Acid (DA)** as the main toxin. Cultivated **blue mussels** sampled during the first outbreak of **ASP** in Canada (at eastern Prince Edward Island) during 1987 contained up to 790 *mg* **DA**/g whole mussel wet tissue and up to 1,280 and 1,500 *mg/g* soft tissue and digestive gland, respectively. During August-October 1988, **DA** was also detected in **soft-shell clams** from the southwest Bay of Fundy, Canada. In October 1991, **DA** was detected in **razor clams** from Oregon and Washington States (US).

(ii) Epidemiology:

DA accumulates in planktivorous fish and shellfish, resulting in **trophic transfer** to marine mammals and humans. **DA** related ASP was first discovered in 1987, when blue mussels grown off Prince Edward Island, Canada became toxic after a long-lasting bloom of *Pseudo-nitzschia multiseries* (Fig. 3.1 c). It caused three deaths and 105 cases of acute human poisoning following the consumption of **blue mussels**. Cases of outbreaks have also been reported from Europe, Central and South America, Australia, Tasmania and New Zealand. Human toxicity has been reported after ingestion of 1.0 –5.0 *mg/kg* **DA**.

> **BOX 3.2**
> **OTHER MARINE ORGANISMS CONTAINING ASP/DA TOXINS**
>
> **DA** has also been detected in the viscera of **Dungeness crabs** (*Metacarcinus magister*) from Washington and Oregon States (US) [pl. see Chapter - 4]. These crabs are supposed to be opportunistic predator-scavengers at benthos, preying on toxic subtidal razor clams. In September 1991, **brown pelicans** (*Pelecanus occidentalis*) and **cormorants** (*Phalacrocorax penicillatus*) died in Monterey Bay, California, after eating **anchovies** (*Engraulis mordax*) contaminated with **DA**. Frequent *Pseudo-nitzschia* blooms along the west coast of United States have resulted in repeated exposure of marine mammals to **DA**, causing mortality, especially of **California sea lions** (*Zalophus californianus*), over 400 of them dying during May and June 1998.

(iii) The toxins and their source:

DA, the chief naturally occurring toxin responsible for **ASP/DAP**, is actually a crystalline water-soluble **acidic amino acid**, belonging to the **'kainoid'* class of compounds, isolated from a variety of marine sources including macro- and microalgae. It was originally isolated from a **Red macroalga** (*Chondria armata*) in 1950s. For centuries, Japanese used this alga for the treatment of roundworm disease and as an insecticide. Two decades later, **DA** was detected in another **Mediterranean red macroalga** (*Alsidium corallinum*). The diatom, *Amphora coffaeformis*, is also known to produce **DA**.

As many as 12 species of *Pseudo-nitzschia* from different parts of the world have been found to be the frequent source of **DA**, *P. multiseries* and *P. australis* being the most toxic ones. Upwelling of coldwater with high nitrogen concentrations stimulates the increase in the population of this diatom. Other external factors like silicon and phosphorus limitation trigger DA production in *Pseudo-nitzschia*. The peak bloom of this diatom was noticed in Canada in 1987 after an unusually dry spell in late summer, followed by severe rainstorm in early September. In September 1991, deaths of **pelicans** and **cormorants** in Monterey Bay, California were attributed to an outbreak of the **DAP** produced by *P. australis*. The latter was consumed by fishes (Anchovies) that in turn were eaten by the birds (BOX 3.2). In the **blue mussel** and the **oyster** (*Crassostrea virginica*) [PLATE IV g] **DA** has been found accumulated in the gut. Anatomical **DA** distribution was studied in **scallops** (*Pecten maximus*) [PLATE IV h] and substantial amount of **DA** was found routinely in their digestive glands.

> *The *kainoids* are a class of **non-proteinogenic, excitatory** and **excitotoxic pyrrolidine dicarboxylates.** They act as **glutamate receptor agonists** by activating ionotropic glutamate receptors. The parent compound, *α-kainic acid*, isolated from a **red alga** (*Diginea simplex*), is used in Asian countries for treatment of intestinal worms in children.

(iv) Effect of toxin and Symptoms:

DA is a potent **glutaminergic excitatory neurotoxin**, targeting the hippocampal and brain stem regions of the central nervous system, the areas involved in processing memory and visceral function, respectively. **DA** acts on excitatory amino acid (*L-glutamate* and *L-aspartate*) receptors *viz.*, †*N-methyl-D-aspartate* (NMDA) and †*kainate receptors* and thus on synaptic transmission.

> †**Kainate receptors** (KARs) are ionotropic receptors that respond to glutamate neurotransmitter. They were first identified as a distinct receptor type through their selective activation by the **agonist kainate**, a drug first isolated from the **red alga** (*Digenea simplex*). **Postsynaptic kainate receptors** are involved in excitatory neurotransmission. **Presynaptic kainate receptors** have been implicated in inhibitory neurotransmission by modulating release of the inhibitory neurotransmitter, **GABA**, through a presynaptic mechanism.

Since **DA** is a **glutamate analogue**, it binds with great affinity to glutamate receptors. **Glutamate** and also **NMDA** subclass opens membrane channels permeable to Na^+, leading to Na^+ influx and membrane depolarization.

The symptoms of **DA-poisoning** appear within 15 *minutes* to 38 *h* after mussel-consumption. As **DA** is potentially **neurotoxic** to both the central and peripheral nervous systems, it becomes emetic, causing gagging and vomiting, likely through its effect on the vomit centre in the brain. It produces a syndrome of axonal sensory-motor neuropathy, **amnesia** (memory loss), seizures, coma and death. Because of its impact on 'memory', **DA** intoxication has been aptly called as *__Amnesic Shellfish Poisoning__ (ASP). Other most common symptoms are abdominal cramps, headache and diarrhoea.

> ***Amnesic** = a person experiencing a partial or total loss of memory.

(v) Treatment:

Symptomatic and supportive. It has been observed that the seizures respond to intravenous infusion of **diazepam** and **phenobarbital**.

(vi) Prophylaxis:

- **Detoxification:**

Till date there have been no useful methods devised for effectively reducing phycotoxins in contaminated shellfish. All methods tested have been unsafe, too slow or economically unfeasible or have yielded products unacceptable in appearance and taste. Mussels were reported to take up **DA** rapidly but also depurated rapidly, while other bivalves retained **DA** for longer periods.

- **Prevention:**

Extensive monitoring of the marine environment and the contaminated fishery products together with regulations will be required to prevent shellfish poisoning incidents. Harvested fishery products containing too much toxin are required to be destroyed.

- **Regulations:**

In Member States of the European Union, a guideline value of 20 *mg/kg* is valid for the total **ASP** toxin content in the edible parts of molluscs. Fishery product harvesting areas are closed when toxin level in shellfish exceed the guideline value.

3.1.4 Neurologic Shellfish Poisoning (NSP) or Brevetoxin Poisoning:
(i) Shellfish:
Oysters (*Crassostrea gigas*), **Cockles** (*Austrovenus stutchburyi*), **Green shell mussels** (*Perna canaliculus*) and **Clams** (*Chione cancellata* and *Mercenaria* spp.) [PLATE IV] are found associated with **NSP**.

(ii) Epidemiology:
Until 1992/1993, **NSP** was endemic to the Gulf of Mexico and the east coast of Florida, where Red Tides by heavy blooms of dinoflagellates were reported associated with NSP way back in 1844. The largest and best documented outbreak of **NSP** in United States occurred in North Carolina. A prominent outbreak of **NSP** occurred in New Zealand in 1992–1993 with over 180 cases reported over a period of several weeks due to consumption of **cockles**, **green shell mussels** and **oysters**. In 1995-1996 summers, a **NSP-associated** *aerosol toxin problem (see ahead) was reported from South Africa (False Bay, a coastal resort of Hermanus in Walker Bay).

(iii) The toxins and their source:
In 1947, blooming of a **dinoflagellate** (*Gymnodinium breve* or *Ptychodiscus breve*; renamed as *Karenia brevis* since 2000) [PLATE III e] was identified as the sole agent responsible for all the outbreaks described since 1844. Blooms of *K. brevis* cause Red Tides (or HABs) annually throughout the Gulf of Mexico; but also in the West Atlantic, Spain, Portugal, Greece, Japan and New Zealand. In October 1987, the same dinoglagellate bloom became entrained in the Gulf Stream off eastern Florida.

A tasteless, odourless, heat and acid stable, lipid-soluble **cyclic polyether brevetoxin** (BTX) produced by *K. brevis* is now known as the main causative agent of **NSP**. The body of *K. brevis* is fragile (*i.e.*, naked, having no outer shell of polysaccharide plates like in other dinoflagellates) and is readily broken down due to wave action along beaches, thus releasing a **toxic '*aerosol'**. During in-shore red tide, the **toxic aerosol** contains the toxins and fragments of the organism both in the droplets and those attached to salt particles. **BTXs** can be highly concentrated in the aerosol of sea spray generated by waves hitting the shore during a Red Tide.

***Aerosols** = extremely small solid particles or very small liquid droplets, suspended in the atmosphere.

Any detectable level of **BTX**/100 *g* shellfish tissue is considered potentially unsafe for human consumption. In addition to **BTX**, some phosphorus containing **ichthyotoxic** compounds resembling **anticholinesterases** and **hemolytic toxins** have also been isolated from *K. brevis*.

In 2000, a bloom of another small Raphidophycean **flagellated unicellular alga** (*Chattonella verruculosa*) found in Delaware bays and creeks (estuary zone of Delaware River on the northeast seaboard, USA), was responsible for massive **fish kills** (Clupeoid menhadens). This bloom contained elevated levels of **BTXs**, although neither *K. brevis* nor *K. brevis*-like species were found involved.

(iv) Effect of toxins and Symptoms:
Being **neurotoxic**, BTXs activate voltage-sensitive Na^+ channels, thus causing sodium influx and nerve membrane depolarization. Respiratory problems associated with the inhalation of aerosolized BTXs are due in part to opening of Na^+ channels. The symptoms in non-asthmatic persons usually end rapidly within a few hundred feet of the seashore or upon entering air-conditioned cars or houses.

Since **BTXs** are also **lysosomal proteinase inhibitors** (= cathepsins found in phagocytic macrophages and lymphocytes), acute and chronic immunologic effects (including the release of inflammatory mediators that culminate in fatal toxic shock) may be associated with exposure to aerosolized **BTXs**.

Massive fish kills during Florida Red Tides were mainly due to exposure to neurotoxic BTXs along with a possible hemolytic action. **BTX**-associated mortality was postulated in **Bottlenose Dolphins** (*Tursiops truncatus*) in southwest Florida in 1946 and 1947.

The **symptoms** appear within 30 *minutes* to 3 *h* and last for a few days, the chief clinical symptoms being an acute gastroenteritis with neurologic symptoms and a reversible respiratory syndrome following inhalation of aerosolized red tide toxins. Since **BTXs** stimulate post-ganglionic cholinergic fibres, they result in respiratory irritation, conjunctival irritation, copious catarrhal exudates, rhinorrhea, non-productive cough and broncho-constriction. More sysmptoms include nausea, vomiting, diarrhea, rapid reduction in respiratory rate, cardiac conduction disturbances, chills, sweats, reversal of temperature, hypotension, arrhythmias, numbness, tingling, paresthesias of lips, face and extremities, cramps, paralysis, seizures and coma. No mortality or chronic symptoms are reported. Some people also suffer from the symptoms like dizziness, tunnel vision and skin rashes.

(v) Treatment:

Treatment is primarily supportive. Fluid replacement, observation of respiratory functions and the administration of sedatives and pain mitigation are the main remedies, as there is no specific antidote available for BTXs. Gastrointestinal decontamination with activated charcoal, for patients diagnosed within the first 4 *h* of post ingestion, has been recommended. **Mannitol** (the recommended treatment in ciguatera) may be useful in early treatment. A recently discovered natural antagonist of BTX, known as **brevenal**, produced by *K. brevis* has proved to have therapeutic value in the treatment of NSP.

(vi) Prophylaxis:

- **Detoxification:**

The most usual way of depurating bivalves is self-depuration, achieved by moving shellfish stock to clear water. Cooking and freezing is ineffective. One of the most promising agents discovered is **ozone** which assists in the depuration of mussel tissues.

- **Prevention:**

Monitoring programmes relying on microscopic identification and enumeration of harmful dinoflagellate/algal taxa in water samples generally suffice for preventing human intoxication. In case of respiratory irritation due to aerosolized Red Tide toxins, particle **filter masks** are used or retreat to air conditioned environment provides relief from the airborne irritation.

- **Regulations:**

In Europe, NSP-producing algae are monitored and fishery product harvesting areas are closed after the simultaneous presence of algae in water and toxin in mussels is noticed. Since mid 1970s, the **Florida Department of Environmental Protection** has conducted a control program with the closure of shellfish beds when *K. brevis* concentrations were greater than 5000 cells/*l*, until 2 weeks. This prevented cases of ingestion of NSP - contaminated shellfish but not the respiratory irritation associated with exposure to aerosolized Red Tide toxins. **The Florida Poison Information Center** at the University of Miami initiated a toll free 24 *h*/day Marine Hotline in 1997 to enhance reporting of marine toxin associated diseases. The **FDA** has established an action level of 0.8 *ppm* **BTX** equivalents.

3.1.5 Azaspiracid Shellfish Poisoning (AZP):

(i) Shellfish:

Irish blue mussel (*Mytilus edulis*) [PLATE IV a] is the chief causative agent of **AZP**. It is a widely distributed and dominant species in the intertidal zones throughout temperate waters, where there is wave-exposed rocky substrate available for the mussels to remain fastened. Other bivalves implicated with **AZP** are the **Oysters** (*Crassostrea gigas*), **Scallops** (*Pecten maximus*) and **Cockles** (*Cardium edule*) [PLATE IV g-i].

(ii) Epidemiology:

It was in 1995 when a good number of human poisoning cases were found to be due to the consumption of cultivated **blue mussels** from Killary Harbour, Ireland. Cases of contamination recurred in 1997 and caused human intoxication in Arranmore Island region of Donegal, Northwest Ireland and other European countries. In August 2000, similar incidents of food poisoning were reported from some regions of England (Sheffield, Warrington, Alyesbury and the Isle of Wight)

after consumption of processed mussels originating from the SW coast of Ireland. In 2008, two cases of **AZP** were reported in US and linked to consumption of an imported mussel product from Ireland.

(iii) The toxins and their source:

Earlier, a marine dinoflagellate (*Protoperidinium crassipes*) was supposed to be the progenitor of the toxin named as **Azaspiracid** (AZA), but recently another dinoflagellate *Azadinium spinosum* (PLATE III f) has been identified as the chief AZA-producing organism. The first report of

AZA in Asia, is from a **sponge** (*Echinoclathria sp.*) found in Japan. From historical importance point of view, originally this biotoxin was named as '**Killary-toxin**' or **KT-3,** after the name of a Killary harbour, in Ireland.

Typically, these **toxins** are found accumulated in the **digestive gland** (hepatopancreas). The highest levels of total azaspiracids recorded are in mussels, followed by oysters, clams, scallops and cockles. Maximum toxin levels are recorded during late summers.

The AZA is a **nitrogen-containing polyether** with a unique **tri-*spiro* ring-assembly**, a **heterocyclic amine** (piperidine or *azacycloalkane* or '*aza*' group) and an aliphatic carboxylic *acid* group, thus becoming the basis of the name of the biotoxin '*AZA-SPIR-ACID*' (**AZA**). Over the last decades, more than 30 analogues of AZA have been identified in shellfish from coastal regions of western Europe, as well as NW Africa and eastern Canada. **AZA$_1$** has been considered as the chief azaspiracid and its structure ($C_{47}H_{71}NO_{12}$,) was devised in 1998 after its successful isolation from Irish blue mussel (*Mytilus edulis*). At physiological pH, **AZA$_1$** exists as a zwitterion (*i.e.*, containing both positive and negative charges) and confers detergent-like properties to the molecule.

(iv) Effect of toxins:

Gastrointestinal disorders have commonly been associated with **AZP** but **neurotoxic effects** are also observed in mouse bioassays. It is a unique toxin that targets liver, lung, pancreas, thymus, spleen (T- and B-lymphocytes) and the digestive tract. It has been revealed that **AZA$_1$** is cytotoxic to various cell types and cytotoxic effects are both time- and concentration-dependent. However, **AZA$_1$** takes an unusually long time (>24 *h*) to cause complete cytotoxicity in most cell types. **AZA$_4$** appeared inhibitor of plasma membrane Ca^{2+} channels, affecting at least store-operated channels in Ca^{2+} signaling, showing an effect clearly different from other **AZA** analogues. Recently it has also been demonstrated that **AZA$_1$** acts as a potent **teratogen** to fish.

The **Lowest-Observable-Adverse-Effect-Level** (**LOAEL**) of **AZA** has been estimated at 1.4 *μg/g* of raw mussel meat. A **No-Observable-Adverse-Effect-Level** (**NOAEL**) is also estimated at 1/10th of the **LOAEL** and is used to estimate the maximum permitted levels in food. Thus, the **NOAEL** would be about 14 to 42 *μg*/person (@ consumption of 100 to 300 *g* shellfish meat/meal).

(v) Symptoms:

Similar to DSP toxins, human consumption of AZA-contaminated shellfish results into acute symptoms of nausea, vomiting, diarrhea and stomach cramps. These symptoms may persist for about 2–3 days. It has been concluded that **AZAs** may contribute to **chronic disorders** in the GI tract (*i.e.,* Crohn's disease, ulcerative colitis) or cancers of stomach, intestines and colon.

(vi) Treatment:

Similar to earlier poisonings but still not properly understood.

(vii) Prophylaxis:

- **Detoxification:**

In winters, when shellfish are free of contamination by DSP toxins, AZP toxins may occur in mussels. The long duration of toxicity periods, which often extend to nearly six months, is troublesome. During the initial stages of intoxication, digestive glands of mussels contain most of the AZP toxins. This unusual distribution of AZP toxins within the shellfish tissue leads to slow rate of natural depuration.

- **Regulations:**

As a consequence of risk assessment, the first regulatory limit established in Ireland in 2001 for AZAs was 100 *μg/kg* shellfish meat. The EU Commission introduced a provisional limit for AZAs as 160 *μg/kg* shellfish meat.

3.1.6 Ciguatera Shellfish Poisoning (CSP):

The filter-feeding **giant clams** (*Tridacna maxima*) [PLATE IV f], frequently consumed as food in the South Pacific, are often found involved in **Ciguatera-like Fish Poisoning** (CFP) incidents, usually typical of the ingestion of toxin-containing tropical coral reefs fishes. As explained earlier (Chapter – 1), the main causative agent of CFP is a benthic dinoflagellate (*Gambierdiscus* spp.) [PLATE III b]. A family of **polyether neurotoxins** (CTXs) produced by this dinoflagellate potentially accumulate through the food chain and attain high concentrations in fishes at the upper trophic levels, thus posing serious health risks to the consumers.

A number of toxicological investigations conducted on the Pacific islands *viz.,* Lifou (Loyalty Islands, New Caledonia), Raivavae (Austral Archipelago, French Polynesia) and Emao (Republic of Vanuatu); have revealed a new link between the presence of **cyanobacterial blooms** and the occurrence of CFP in giant clams and/or fish from lower trophic levels. Akin to dinoflagellates, **Cyanobacterial toxins** (ciguatoxins-like compounds) enter the food chain directly *via* **parrotfishes** (*Scarus* sp.) which graze on cyanobacterial mats or *via* molluscivorous **long-nosed emperor fishes** (*Lethrinus* sp.) which prey on contaminated molluscs.

Based on the presence of a **benthic cyanobacterium**, *Lyngbya majuscule*, in the guts of a large number of poisonous fishes, Halstead (1967) hypothesized that these cyanobacteria might serve as a primary source of CTXs or their progenitors. Other benthic cyanobacteria associated with seafood poisoning are *Hydrocoleum, Anabaena, Phormidium, Spirulina, Leptolyngbya* and *Oscillatoria* spp. When cyanobacterial blooming takes place, there are chances of filter-feeding giant clams of becoming contaminated, thus providing a new link for the transfer of **cyanotoxins** to upper trophic levels including humans. The name **Ciguatera Shellfish Poisoning** (CSP) has, thus, been proposed to designate this **new kind of toxicity**.

The **symptoms** of this particular poisoning include the characteristics of typical CFP (reversal of sensations, itching and bradycardia) associated with additional symptoms like the burning of the mouth and the throat that appear very quickly and are followed by severe paralysis.

NOTE: For mode of action and symptoms etc. of CTX-poisoning, please refer to CHAPTER – 1.

3.2 GASTROPOD SHELLFISH-POISONING:

Constituting about 80% of the Molluscans, the univalved **Gastropods** (Abalones, Snails, Slugs, Limpets, Sea Slugs, Whelks, Conchs etc) have an extraordinary diversification of habits and habitats, ranging from gardens, woodland, deserts, mountains, small ditches, great rivers and lakes; estuaries; coastal areas to ocean depths. As per **feeding habits**, some scrape algae from rocks along the river/pond/ocean floor and some cling to large freshwater/marine plants and feed on them. Interestingly, **parasitism** and **carnivorous/predatory/necrophagous** [= feeding on corpses or carrion] life styles are common in marine gastropods. *Asterophila* (Eulimidae) from Antarctic Peninsula, is an exclusively **endoparasitic** gastropod, forming cysts in the arms and central discs of asteroid sea stars. Moreover, certain slugs and terrestrial snails participate in transmission of digenetic Trematode/Nematode larval stages, including those pathogenic to man. On the other hand, **venomosity** among gastropods (see ahead) is also not an exception.

Marine gastropods are consumed worldwide due to their **nutritional quality** and economic value in international market. However, like fishes and bivalve molluscs, the Gastropods are not an exception to becoming **highly toxic** due mostly to the **'exogenous toxic sources'**, ranging from bacteria to dinoflagellates/algae; but sometimes through **'biosynthetic pathway' (endogenous)**, too.

Toxic gastropods mostly include - Horse Conchs (Fasciolariidae), Nassa mud snails or dog whelks (Nassariidae), Moon snails or Necklace shells (Naticidae), Olive shell snails (Olividae), Murex snails or Rock snails (Muricidae), Whelks or True whelks (Buccinidae), Pear whelks (Busyconidae), Turban snails (Turbinidae), Frog snails or Frog shells (Bursidae), Triton's trumpet or Triton snail (Charoniidae), Babylon Snails (Babyloniidae), Top shell (Tegulidae), Volutes (Volutidae), Florida fighting conch (Strombidae), Oregon hairy Triton (Cymatiidae); Crown Conchs (Melongenidae) *etc* (PLATE V).

As for the **'exogenous poisoning'** aspect, **carnivorous gastropods** become **highly toxic** *via* feeding on bivalve mollusks and pose health hazards to human beings who utilize them as food. On the basis of prime symptoms, these **'poisonings'** are identified by the same name as used for describing them in 'fishes' as well as 'bivalve molluscs' *e.g.,* **Ttetrodotoxin-Poisoning** (TTX-poisoning), **Ciguatera Poisoning** (CTX-poisoning), **Paralytic Shellfish Poisoning** (PSP-poisoning), **Diarrhoeic Shellfish Poisoning** (DSP-poisoning) and **Neurotoxic Shellfish Poisoning** (NSP-poisoning).

PLATE V

Examples of Gastropod shellfish-poisoning: **(a)** Horse Conch (Fasciolariidae); **(b)** Nassa mud snails or dog whelks (Nassariidae); **(c)** Moon snails or Necklace shells (Naticidae); **(d)** Olive shell snails (Olividae); **(e)** Murex snails or Rock snails (Muricidae); **(f)** Whelks or True whelks (Buccinidae); **(g)** Pear whelks (Busyconidae); **(h)** Turban snails (Turbinidae); **(i)** Frog snails or Frog shells (Bursidae); **(j)** Triton's trumpet or Triton snail (Charoniidae); **(k)** Babylon Snails (Babyloniidae); **(l)** Top shell (Tegulidae); **(m)** Volutes (Volutidae); **(n)** Florida fighting conch (Strombidae), **(o)** Oregon hairy Triton (Cymatiidae); **(p)** Crown Conchs (Melongenidae).

Except TTX-poisoning, which is **'bacterial'** in origin, all other cases are caused *via* certain toxic **dinoflagellates/algae** used as food by the bivalve molluscans or *via* dinoflagellate-generated toxins being accumulated in the body of preys utilized by cephalopods or crustaceans.

In case of **'endogenous biosynthetic pathway'**, the poisoning from a heat stable metabolite **'Tetramine'**, found in the salivary glands, is exclusive to edible Gastropods.

3.2.1 Ttetrodotoxin-Poisoning:

As is well-known, **TTXs** are either produced by **symbiotic bacteria** (mainly *Vibrio* spp.) or that they are exogenously accumulated through the diet *via* dinoflagellate blooms (*e.g., Alexandrium tamarense* or *Prorocentrum cordatum*) [PLATE III c & Fig. 3.1 d]. More specifically, TTXs and its analogues are found in the toxic **Nasssariid Gastropods** *viz., Nassarius* (= *Alectricon*), *Niotha,* spp. etc. In Asia, Nassariids are considered a delicacy and often cooked by boiling or frying. Consumption of these marine Gastropods sometimes leads to poisoning incidents as reported in Asian (like China, Vietnam, Taiwan, Japan) and other countries. Besides, members of **carnivorous/predatory**, **Trumpet shell** or **Tritons** (Charoniidae), **Frog Shell** (Bursidae), **Moon Snails** (Naticidae), **Olive Snails** (Olividae) and **necrophagous Ivory Shells** (Babyloniidae) etc. (PLATE V) have also been found implicated with TTX poisoning.

A starfish has been found the source of intoxication to some gastropods; as fragments of the **starfish** (*Astropecten polyacanthus*), found in the digestive tract of the carnivorous **trumpet shell** (*Charonia* sp.), showed the presence of TTX as the toxic molecule. As for puffers, **TTX** is accumulated in specific organs, like **digestive gland** and **muscles** of these gastropods.

Interestingly, pufferfish ingest/accumulate TTX from necrophagous small or medium-sized marine snails, while on the other hand, these snails ingest/accumulate the toxin from dead pufferfish. Thus, TTX produced by bacteria gets transferred to higher organisms *via* food chain.

Symptoms to the victims are typically similar to those of the TTX puffer-fish poisoning *e.g.,* hypertension and other neurological symptoms.

NOTE: For mode of action and symptoms etc. of TTX-poisoning, please refer to CHAPTER – 1.

3.2.2 Ciguatera Poisoning:

As explained earlier (Chapter – 1), **Ciguatera Fish Poisoning** (CFP) is the most prevalent **non-bacterial** food-borne form of poisoning reported from French Polynesia, developing after the consumption of coral reef fish, contaminated with **ciguatoxins** (CTXs) produced by some dinoflagellates (*Gambierdiscus*) [PLATE III]. Globally, the **endemic regions** are mostly from the coral rich areas of the Caribbean Sea, Pacific and Indian Oceans where the climate and water conditions are most favourable for the blooming of *Gambierdiscus* spp. Other dinoflagellates *viz., Prorocentrum* spp., *Gymnodinium sangieneum* and *Gonyaulax polyedra* are also implicated with ciguatera poisoning. Natural and anthropogenic disturbances are the main factors likely to worsen ciguatera risk in coral reef environments.

In June 2014, **Ciguatera Shellfish Poisoning** (CSP) was reported as a major and persistent poisoning following the consumption of the **Gastropod**, *Tectus niloticus* (Tegulidae) from the bay of Nuku Hiva Island (French Polynesia).

As uusual, the **symptoms** of CSP include the characteristics of CFP, (reversal of sensations, itching and bradycardia) followed by severe paralysis.

NOTE: For mode of action and symptoms etc. of CTX-poisoning, please refer to CHAPTER – 1.

3.2.3 Paralytic Shellfish Poisoning (PSP):

As described for bivalve mollusks (mussels), the **dinoflagellates** *e.g., Alexandrium, Gymnodinium* and *Pyrodinium* (PLATE III & Fig. 3.1 b), are known producers of **PSP toxins** *viz.,* **saxitoxins** (STXs) and **gonyautoxins** (GTXs). As a matter of fact, bivalves are used as food by **marine carnivorous gastropods**, thus, accumulating PSP toxins after predation. Consequently, there is always a risk to the consumers who subsist on gastropods for food *e.g.,* those belonging to **Volutidae** (*Zidona* and *Adelomelon* sp.), **Naticidae** (*Natica* spp.), **Olividae** (*Oliva* spp.) [PLATE V]; particularly in the Southeast Asian regions, Argentina, South Taiwan etc. Considerable amount of PSP toxins are found in the viscera and foot muscle of these edible snails, with the dominance of **STXs**, followed by **GTX**.

PSP has also been observed in cases who ingested **whelks** (Buccinidae) [PLATE V f] without removing the salivary glands.

PSP toxins cause paralysis in humans by blocking Na^+ channels in neurons, thereby preventing neurons from functioning normally. Early symptoms include tingling of the lips and tongue, appearing within minutes of eating toxic shellfish. Symptoms may further progress to tingling of fingers and toes and then the loss of muscle control in the arms and legs, followed by difficulty in breathing.

NOTE: For mode of action and symptoms etc. of PSP, please refer to poisonings by bivalves, at 3.1.1.

3.2.4 Diarrhoeic Shellfish Poisoning (DSP):

Similar to bivalve-poisoning, DSP is causative of gastrointestinal disorders after consumption of seafood contaminated with lipophilic biotoxins like **Okadaic Acid** (OA), **Dinophysistoxin** (DTX) and **Yessotoxin** (YTX); produced by a dinoflagellate (*Dinophysis* spp.). When a filter-feeding bivalve feeds on this biotoxin-producing dinoflagellate, the biotoxin gets accumulated in their tissue.

Historically, DSP has not previously been reported in gastropods but very recently OA, DTX and YTX were detected in the flesh and digestive/salivary glands of the gastropod species widely consumed in Korea *viz.*, *Neptunea* (Buccinidae), *Rapana* (Muricidae) and *Turbo* (= *Batillus*) spp. [Turbinidae] [PLATE V]. DSP has also been observed in cases who ingested **whelks** (Buccinidae) without removing the salivary glands.

The main symptoms caused by consumption of DSP-contaminated shellfishes/Gastropods include diarrhoea, vomiting, and abdominal pain. Other symptoms associated with the disease include nausea and headache.

NOTE: For mode of action and symptoms etc. of DSP, please refer to poisonings by bivalves, at 3.1.2.

3.2.5 Neurotoxic Shellfish Poisoning (NSP):

Incidences of **NSP** after consuming gastropods, contaminated by **brevetoxins** (BTXs), are not uncommon. **Horse Conch** (*Triplofusus giganteus*), **Fighting conch** (*Strombus alatus*) and **Pear whelk** (*Fulguropsis spirata*) [PLATE V] are the gastropods known to transfer BTXs to human beings during the **Florida red tide** created by a **dinoflagellate** (*Karenia brevis*) [PLATE III e], frequent along the Gulf of Mexico. As is well known, BTXs are **neurotoxins** which activate voltage-sensitive Na^+ channels, causing Na^+ influx and nerve membrane depolarization. Presence of BTXs has been noticed in the urine samples of the patients diagnosed with NSP, involving gastrointestinal and neurological symptoms *viz.*, nausea and vomiting, paresthesias of mouth, lips and tongue; ataxia, slurred speech and dizziness. Neurological symptoms often lead to partial paralysis and respiratory distress.

NOTE: For mode of action and symptoms etc. of NSP, please refer to poisonings by bivalves, 3.1.4.

3.2.6 Tetramine Poisoning:

A number of poisoning incidents, reported after ingestion of Gastropods, have been found to be due to high levels of an **endogenous toxin**, named *****Tetramine** (*Tetramethylammonium ion*: $(CH_3)_4N^+$), found localized in the **hypobranchial/salivary gland** of 'whelks' or Buccinid gastropods (mostly Red Whelk, *Neptunia antiqua*) [PLATE V f]. Tetramine has also been found in the muscles and viscera of *Neptunea* snails and even in the salivary gland of other marine snails like *Buccinum* and *Busycon* spp. *Neptunea* snails have been the cause of food poisoning in North Atlantic and Northeast Asia regions, especially in Japan; and also in United Kingdom and Denmark. Besides *Neptunea* spp., large amount of tetramine is found in the salivary gland of two other snails *viz.*, **Oregon hairy Triton** (*Fusitriton* sp.) and Crown conchs (*Hemifusus* sp.) [PLATE V o, p].

Tetramine, produced in the salivary gland, is delivered through the ducts into the mouth, from where it acts to paralyze preys, like the bivalves, for which the whelk is a predator. Alternatively, tetramine may be released into the surrounding water, functioning as a defensive substance against potential predators.

*****Tetramine** was first found in the 'beadlet sea anemone' (*Actinia equine*) and later in another 'ball sea anemone' or 'giant Caribbean sea anemone' (*Condylactis gigantean*) and 'Portuguese man-of-war' (*Physalia physalis*).

It is a heat-stable **neurotoxin,** structurally similar to acetylcholine. Therefore, it binds to acetylcholine receptors, thereby acting as a **ganglionic blocking agent** and inhibiting synaptic transmission. It induces long-lasting depolarization blockade in **autonomic nervous systems** and ultimately leading to paralysis of skeletal muscles, similar to †**Curare-like effects**.

Various **symptoms**, attributable to the **ganglion-blocking action** are: visual disturbances, headache, dizziness, sleepiness, abdominal pain, nausea, diarrhoea, ataxia and tingling in the fingers. The symptoms develop within 30 *minutes* after ingestion of snails because of rapid absorption of tetramine from the gastrointestinal tract. The symptoms are generally mild and subside shortly (within 24 *h*) because more than 95% of tetramine is excreted through kidneys.

To prevent tetramine poisoning, it is important to remove salivary glands from live snails.

†**Curare-like effects:** In general, **Curare** is the name given to any highly toxic substance from some plants of **South America**, crude preparations of which are used by indigenous tribes to poison their hunting arrows. **Curare**, is also the name of an **alkaloid drug**, derivatives of which are used in modern medicine, primarily as skeletal muscle relaxants, being administered concomitantly with general anesthesia in certain types of surgeries. These alkaloid toxic substances are stored in the roots, stems and leaves *e.g.*, **strychnine** and **curare** from *Strychnos* sp. and **tubocurarine** from *Chondrodendron* sp.. Functionally, **Curare** is a potential **neurotoxin**, causing muscle paralysis by acting as **acetylcholine antagonist**. It blocks neuromuscular transmission, a process that allows the central nervous system to control the movement of muscles, at the neuromuscular junction, which is the junction between a nerve cell and a muscle cell.

3.3 CEPHALOPOD SHELLFISH-POISONING:

Like fishes, Cephalopod Molluscans (Squids, Cuttlefish and Octopus) (Fig. 3.2) are rich source of Proteins (16%) along with low fat content (0.7 – 1.4%); the rest (about 80%) being water. Usually, it is the mantle, the arms, the ink-sac and liver which are used as food. They are used to be caught for food in the Mediterranean for at least the last 4,000 years. In ancient Greece, Octopus, Cuttlefish and Squid were very common in banquets. Presently, Spain, Italy, China and Japan are major consumers of cephalopods.

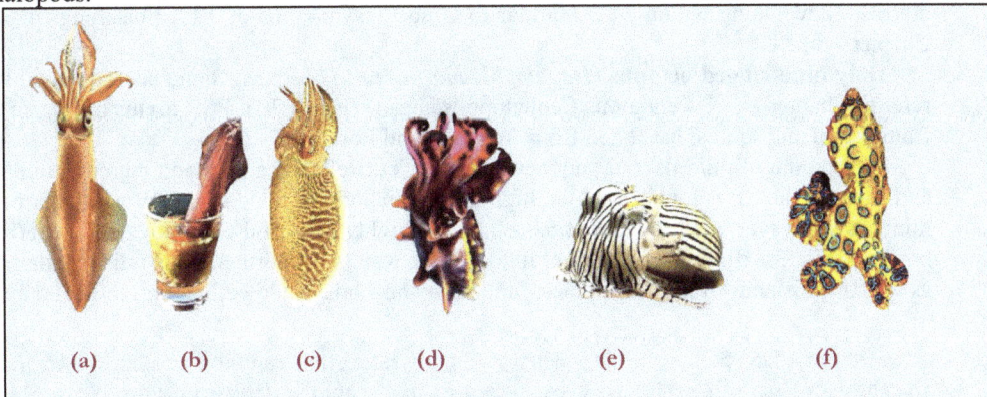

Fig. 3.2 Examples of Cephalopod shellfish-poisoning: **(a)** A Calamari Squid; **(b)** Live squid dipped in a glass; **(c)** A Typical Cuttlefish; **(d)** Flamboyant Cuttlefish; **(e)** Striped pyjama Squid; **(f)** Blue-ringed Octopus.

The **Squids** (Order - Teuthida) constitute the most important seafood accounting for 3.6 million *tons*, followed by *Octopus* and *Sepia* species. China, South Korea, Taiwan, Peru, Japan, Argentina, Thailand and Chile account for about 68% of the squid landings. The Japanese value cephalopod food the most and they consume more octopus/*capita* than elsewhere in the world. The largest producers and exporters of octopus are Morocco, Mauretania and China. Every year, 350,000 *tons* of wild octopus are caught, most of them coming from Asia, particularly China. Thailand, Spain, China, Argentina and Peru are the world's largest exporters of Squids (*Teuthis*) and *Sepia* sp..

Consuming **'raw cephalopods'** is a fascination in Japan, Thailand, China etc. with a belief that raw meat provides more *omega-3 fatty acids* and does not contain any contaminant, supposed to appear during the cooking process. In Thailand, it is unique to watch **'live squid eating'** when a live squid is dipped in a glass filled with a spicy and sour chilli sauce (Fig. 3.2 b).

In addition to being consumed raw, various world cuisines have processed cephalopods by a wide range of culinary techniques such as boiling and steaming, frying, grilling, marinating, smoking, drying and fermenting.

Since cephalopods are usually consumed eviscerated, they are considered at a lower risk of shellfish poisoning (as compared to bivalves) in humans. However, cases of poisoning from cephalopods are often reported. By feeding habit, cephalopods are voracious carnivores and they prey mostly on crustaceans, small fish and other molluscs, which are known vectors of marine phycotoxins. They have been reported to accumulate potential HAB-toxins, *viz.*, **Domoic Acid** (DA), **Saxitoxin** (Chapter – 3, Table 3.1) and **Palytoxin** (BOX 1.4) and, therefore, act as **HAB-toxin vectors**. Thus, cephalopods occupy an important position in the marine food web. **(for HABs, pl. refer to APPENDIX – I).**

Coastal **octopods** and **cuttlefishes** store considerably high levels of **DA** (= Amnesic Shellfish Toxin; pl. refer to 3.1.3) in various tissues, but mainly in the digestive gland, the primary site of intracellular digestion and absorption. They also accumulate **Paralytic Shellfish Toxins** (pl. refer to 3.1.1), in the digestive gland, followed by kidneys, stomach, heart, posterior salivary glands and gills. **Palytoxins** (BOX 1.4) are among the most toxic molecules identified in larger quantities in stranded **octopods**.

Used widely as food in the Mediterranean, Northeast Atlantic, North Pacific (Canada, Japan and China), Southwestern Pacific (Australia and New Zealand); **squids** (*Loligo*, *Todarodes* and *Sepioteuthis* spp.) are called *calamari* (< Spanish, *Calamar*, Italian *Calamaro* for squid), in English-speaking countries (Fig. 3.2 a). After consuming uncooked or undercooked squids, they have been reported to cause food-poisoning, called **vibrio illness**, caused by Gram-negative basophilic *Vibrio* bacteria (TTX producers, as in case of **'puffer-poisoning'**; pl. refer to 1.3.2, Chapter - 1). The main symptoms of **'calamari food poisoning'** include *digestive disturbances *viz.*, nausea, vomiting and diarrhea, developing within 24 *h* (similar to Grade 1 symptoms of TTX poisoning; pl. refer to 1.3.2, Chapter – 1).

Only **blue-ringed octopus** (*Hapalochlaena fasciata*) has been found contaminated with **TTX** (pl. refer to Chapter – 1 & Venomous Cephalopods ahead) (Fig. 3.2 f). **PSP toxins** have also been reported from octopi and squids, harvested from Australia and Portugal.

On account of mild flavour and meaty texture, **cuttlefish** is a common ingredient in Mediterranean and Asian dishes. It is low in total fat, high in protein and is a good source of a number of vitamins and minerals. However, the little (about 6.0 - 8.0 *cm*) highly colourful **Flamboyant Cuttlefish** (*Metasepia pfefferi*) [Fig. 3.2 d], found in tropical Indo-Pacific waters off northern Australia, southern New Guinea as well as in islands of the Philippines, Indonesia and Malaysia, dwelling on sandy and muddy

> *Apart from food-poisoning, digestive disturbances are often caused by eating raw or undercooked Cephalopods (or fish) which have been parasitized by **anisakid larvae** (= **anisakiasis**). Most cases of *anisakiasis* occur in Japan but cases are not uncommon in Western countries. *Anisakis* is a **Nematode** parasite which lives in the stomach of marine mammals like dolphins and sperm whales. While completing the life cycle, its larvae utilize small crustaceans as their host from where they seek their entry into the fish or cephalopods, feeding on them. To be free from the impact of such harmful parasitism, the cephalopods are recommended to be cooked at 70°C for about 2.0 *minutes*.

substrates, is an exception to **edibility**. These beautiful cuttlefish get their name from the flamboyant pink, yellow and black ripples they make with their bodies when alarmed. They are the only species of cuttlefish known to carry a unique **toxin** in their muscles. Due to highly **toxic muscles** (with unusual acids), it has been found poisonous to eat and it has shown that the toxin is as lethal as that of the Blue-ringed Octopus.

To be free from all such poisonings, the best suggested method is to freeze cephalopods at -20ºC. This strategy is also effective for killing parasites.

NOTE: For mode of action and symptoms etc. of DA-poisoning and PSP, please refer to poisonings by bivalves earlier in this Chapter and for TTX-poisoning in Chapter -1.

3.4 VENOMOUS MOLLUSCS:

Among Molluscans, venom glands along with venom-injecting device are diagnostic to some marine **Gastropods** and **Cephalopods**.

3.4.1 Venomous Gastropods:

Gastropoda is the second largest class (after the Insecta), with about 13,00 Genera and 40,000–90,000 living species of snails and slugs. Since they utilize a variety of ecosystems, evolutionarily they are regarded as the most successful group. Besides living in terrestrial ecosystems (deserts, mountains, backyards and beaches), they are found in ocean depths, freshwater lakes and streams. The **'venomous'** Gastropods belong to about 20 Families of marine carnivorous Order (Clade) **Neogastropoda** under Subclass - Caenogastropoda (= **Prosobranchia**), including the examples like **Whelks** (Buccinidae), **Muricid Rock Snails** (Muricidae), **Volutes** (Volutidae), **Harps** (Harpidae), **Cones** (Conidae), **Vampire Snails** (Colubrariidac), **Turrid Sea snails** (Turridac), **Auger Snails** (Terebridae) etc. [pl. refer to PLATE V e,f & m for Muricids, Buccinids, and Volutids and Fig. 3.3 for the rest].

Fig. 3.3 Some examples of venomous Gastropods: **(a)** Cone Shell (Conidae); **(b)** Vampire snail (Colubrariidae); **(c)** Turrid snail (Turridae); **(d)** Auger Snail (Terebridae).

By feeding habit, they are herbivores, detritivores, scavengers, carnivorous predators and even parasitic. For obtaining food, all gastropods have a specialized ribbon-like structure, called a **radula** (so-called tongue) covered with rows of **chitinous teeth** (*e.g.,* taxoglossates with 2 teeth in a row). Possessing a highly specialized **'radula'**, the venomous gastropods were ealier placed under Superfamily **Toxoglossa** (Gk. *toxikon* = poison; for use on arrows + *glossa* = tongue) *i.e.,* the carnivorous marine snails where the **'radula'** (= tongue) works like poison fangs (= Toxoglossate Radula); the **large teeth** (radular teeth/harpoons) being reduced in number and becoming perforated. These teeth are connected with a large **poison gland** in the esophagus through slender ducts. Because some gastropods feed on blood suctorially (hematophagous), they have a minute radula.

Some **important examples** of venomous snails are as follows:
- **CONE SHELLS:**

They include the **most venomous gastropods** from **Conidae** (*Conus* spp.) found in the tropical and subtropical seas, at depths ranging from the sublittoral to 1,000 *m*. The shell is many-whorled, inverted cone-shaped or oval and quite variable in the tuberculations on the spire and body whorl, striae and the colour pattern (Fig. 3.3 a, 3.5 a). *Conus* has unusually high species diversity with about 706 valid species. They are nocturnal, active predators, dwelling in tidal waters under rocks or in coral reefs. During the day, they lay buried under the sand. Depending on the target prey, cone snails are categorized into three groups *viz.,* the **vermivores,** feeding on polychaetes, hemichordates and echiuroid worms, the **molluscivores** hunting on other gastropods and the **piscivores** hunting on fish.

The highly specialised **venom apparatus** of *Conus* spp. is composed of a **venom gland** (= bulb), **salivary glands** (and accessory salivary glands), **radular sac, pharynx, proboscis** and a **radula** (Fig. 3.4). The latter acts as a delivery system and is hollowed and barbed to resemble a miniature **harpoon**. While the cone snail is hunting, a single harpoon is loaded from the radular sac to the tip of the long and **extensile proboscis** (Fig. 3.5a). Once the cone snail uses a harpoon to inject venom into its prey, it is discarded. Prior to another strike, a new harpoon is reloaded from the radular sac.

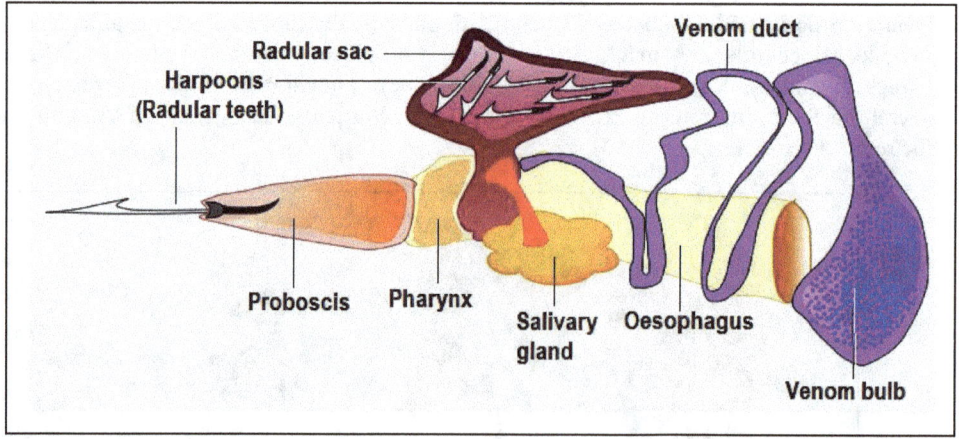

Fig. 3.4: **Venom apparatus** of *Conus* spp.

When the cone snail senses waterborne chemical signals *via* a specialized **chemoreceptor organ** (the osphradium), searching of the prey begins with the extensile proboscis where, in its lumen, a single dart-like radular tooth (harpoon), loaded from the radular sac, is tightly held by circular muscles and filled with venom. When the tip of the proboscis comes in contact with the target, the radula is rapidly thrust into the prey and acts like a hypodermic needle to inject the venom (Fig. 3.5 a). This radular tooth also serves as a harpoon to bring the immobilized prey back to the mouth to swallow it entire, as exemplified by **piscivorous species** (*Conus geographus*).

Biosynthesis and delivery of a cocktail or a complex mixture of of **cysteine-rich, Conantokins/ Conotoxins** (Conopeptides; small disulfide bonded peptides) takes place at the venom apparatus site. The epithelial cells bordering the venom duct are the main site of **conotoxin** production. The muscular **venom bulb** undergoes rapid contractions for the circulation of the venom inside the duct up to the pharynx, where conotoxins undergo sorting and maturation.

To date, approximately 2,000 toxins from about >70,000 **bioactive helical peptides** have been identified in *Conus* spp. Various **Conantokins** (< Filipino, *antokin* = sleepy) are named after the *Conus* species from which they are derived *e.g.,* **Conantokin-G,** also known as **'sleeper peptide'** is a small peptide isolated from the piscivorous species, *C. geographus,* **Conantokin-R** from *C. radiatus* etc.

Multiple **neurotoxins** in the venom act as remarkable **selective inhibitors** and **modulators** of **ion channels** (calcium, sodium, potassium), nicotinic acetylcholine receptors, noradrenaline transporters, N-methyl-D-aspartate receptors and neurotensin receptors. By way of **blocking ion-channels** and modulating **neurotransmitter receptors**, they result into paralysis. In the hunting process,

conantokins by way of acting as neurotoxin, put a fish into a **sleeping** or compliant state. When a *Conus* sp. approaches a fish, the latter becomes **'hypnotic'** and then stung with the coneshell's highly lethal neurotoxin.

Recent studies have revealed that the **venom** of *C. geographus* is a variant of fish **'insulin'** (with shorter chains), used to stun the prey. Once this venom reaches fish gills, the fish experiences **'hypoglycaemic shock'**, becomes paralysed to be ingested by the snail.

Fig. 3.5: Act of targeting the prey: (a) *Conus* spp.; (b) Vampire snail.

Most human accidents occur when the snails are picked up by the divers and pressed against a wetsuit or when collectors attempt to clean the shell of a live animal. In general, the venoms of **piscivorous species** (*C. geographus*) are lethal to vertebrates and most of the vermivorous and molluscivorous species are considered harmless to humans. As per an estimate more than 55% of stings from *C. geographus* may be fatal to humans, sometimes leading to death.

In general, the **symptoms** include excruciating burning sensation, numbness and tingling at the wound site. The affected area quickly becomes edematous and erythematous. This frequently spreads over the whole body with pronounced tingling of the oral area. Wound ischemia, muscle paralysis and heart failure may all arise within 6 h of the initial sting. If the patient survives, most systemic symptoms subside within 1–2 days.

No specific **treatment** is reported for Conid stings, but local pressure and immobilization may limit the spread of the toxin *e.g.,* by putting ligature around the affected part. Immersion in hot water and tetanus prophylaxis (like immunization, treatment of wounds and traumatic injuries) also gives relief. Severe cases may require respiratory support.

- **VAMPIRE SNAILS:**

Belonging to Colubrariidae, most of them have a characteristic colour pattern on the shells, with brown spots and streaks on a yellowish or light brownish background (Fig. 3.3 b & 3.5 b), resembling to that of various snakes, hence the name of the family *i.e.,* Colubrariidae. Like those venomous snakes, **Colubrarid Gastropods** are venomous, by way of producing a **cocktail of toxins**, utilized for feeding. Because for their being **hematophagous**, they are better known as **'vampire snails'** *e.g., Cumia* (*Colubraria*) *reticulata* **feeds on the blood** of fish that are resting or sleeping amidst reefs. Thus, they, together with some species of the family Conidae, are the only gastropods capable of **predating vertebrates**.

Both males and females extend their long and thin proboscis more than three times the shell length. The extended proboscis comes in contact with the skin and then gains access to the blood vessels of the fish by the radular action (Fig. 3.5 b). Relative to Conids, the **radula** is minute and its scrapping action creates wounds with blood kept oozing-out. Such adaptations to **hematophagy** involve the use of an **anticoagulant** and a kind of **'venom'** having **anesthetic properties**. The latter contains peptides, similar to **Stichodactyla toxin** (*ShK, ShkT*), a unique K^+ channel blocker obtained from a Caribbean **sea anemone** (*Stichodactyla helianthus*).

In addition to anticoagulants, *C. reticulata* also releases **vasopressive compounds** that increase the blood pressure of fish, so increased as the blood is passively pumped into snail's gut. Another substance, similar to conopeptides, unique to the vampire snails, is **'turritoxin'** (named after 'Turridae', a family of small, deep-sea dwelling snails (Fig. 3.3 c) closely related to Terebrids and cone snails; that can rasp at prey with their radula or stab with detachable needle-like teeth charged with venom). In the hunting process, **'turritoxin'** puts a fish into a **sleeping** or compliant state. This is supported by the fact observed for 'coneshell', that when it approaches a fish, the latter becomes **'hypnotic'** and then stung with the coneshell's highly lethal neurotoxin. Supposedly, vampire snail, too, releases **'turritoxin'** to put its victim into a deep sleep state.

- **AUGER SNAILS:**

The auger snails (Terebridae) are another distinctive group of carnivorous, sand-dwelling gastropods of the tropics, characterized by an elongated shell (Fig. 3.3 d). Their **venom apparatus** is similar to that of *Conus* spp., consisting of a radular sac, a venom duct and venom bulb. Their specialized radula is similarly used as a spear or harpoon to deliver potent **'teretoxins',** analogous to **conopeptide-like neurotoxins**.

- **ROCK SNAILS:**

Rock Snails (Muricidae; PLATE V e) are well-known for the production of a traditionally used reddish-purple natural dye, the ***Tyrian purple*** or ***Phoenician red***, ***Phoenician purple***, ***Royal purple***, ***Imperial purple*** or ***Imperial dye*** (**brominated indoles** and their derivatives) in their hypobranchial gland. In nature, the snails use these secretions as part of their predatory behaviour in order to sedate prey and as an antimicrobial coating on the egg masses. These **indole compounds** are also well-documented as potential **anti-tumor** agents. However, the **Toxic choline esters** *viz.*, **urocanycholine** (murexine) and **senecioylcholine** have also been isolated from their hypobranchial gland. They have been found to have **neuromuscular blocking properties** through inhibition of nicotinic acetylcholine receptors. Various toxicity studies have verified that the salivary gland extracts act as a depressant on the central nervous system, causing vasodilation, hypotension and bradycardia.

3.4.2 Venomous Cephalopods:

Octopi and **Cuttlefish** represent the potential examples of **venomous** Cephalopods. From amongst the specific anatomical characters, **salivary glands** have an important role to play in the act of **envenomations**.

- **OCTOPI:**

Despite their small size (12.0 to 20.0 cm), characterized by yellowish skin and blue-black rings, commonly called **'blue-ringed Octopi'** (*Hapalochlaena* sp.) are **highly venomous** (Fig. 3.2 f). They dwell tide pools and amidst coral reefs of the Pacific and Indian Oceans. By habit, on provocation, as a warning display they quickly change colour, becoming bright yellow with 50–60 rings flashing bright iridescent blue within seconds.

Out of four, two species are specially important *viz.*, the **greater blue-ringed octopus** (*H. lunulata*) widespread in tropical and subtropical waters of the Indo-West Pacific and **southern blue-ringed octopus** (*H. maculosa*) from the rocky tide pools along the southern coast of Australia. Their arms are joined at the bases by a web of tissue known as the skirt, at the centre of which lies the mouth. The latter has a pair of sharp **horny parrot-like beaks** and a **radula** for drilling shells and rasping away flesh of the prey, respectively (Fig. 3.6).

Octopods feed on a variety of preys *viz.*, crustaceans, gastropods, bivalves, fishes and birds. The prey is seized with the arms and pulled towards the mouth where the **horny beak** is used to pierce through the tough exoskeleton/scaly covering and releasing the venom.

Out of the **three types of salivary glands** associated with the buccal mass, the submandibular gland contributes to lubricating the passage of the food while the secretion of the **anterior salivary gland** facilitates the action of the viscous secretion of the **posterior salivary glands**. The latter, lying behind the buccal mass (Fig. 3.6 b), is exclusively responsible for the production of a number of different biologically active and **venomous substances**, potential enough to paralyze the prey within a few seconds after capture. The viscous secretions are transported *via* muscular salivary ducts to a common terminal canal opening into the anterior part of buccal cavity, close to the apex of the **salivary papilla** (Fig. 3.6 c).

The **toxic effects** of the secretions from the posterior salivary glands, were first attributed to various biogenic amines *viz.*, tyramine, histamine, acetylcholine, octopamine, dopamine and serotonin; and subsequently to a protein component named **cephalotoxin**. But later on, the posterior salivary glands have been shown to contain a deadly neurotoxin, called **'maculotoxin'**, found pharmacologically quite similar to **Tetrodotoxin** (TTX) and **Saxitoxin** (STX) (as noticed in Puffers).

Studies have also shown that TTX of *Hapalochlaena*, is not an endogenous salivary toxin, and

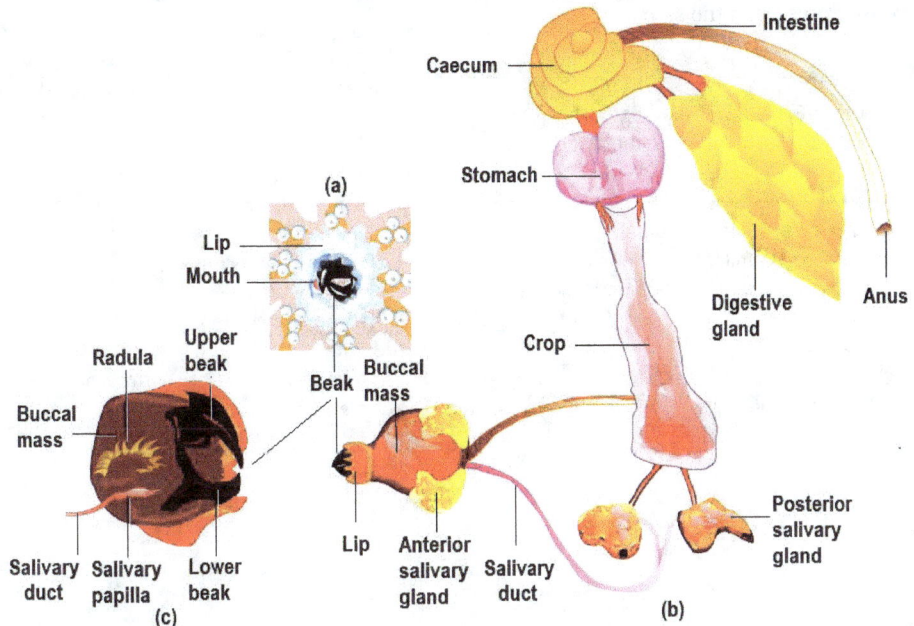

Fig. 3.6: **Digestive system and Salivary glands of an Octopus:** (a) Oral view showing mouth, lip and beak; (b) Digestive sysyem showing the location of salivary glands; (c) Sectional view of buccal mass showing beak, radula and salivary papilla.

supposed to be produced by **endosymbiotic bacteria** harbouring the salivary glands and other parts of the body.

TTX is well-known for **blocking Na$^+$ channels**, leading to **motor paralysis** and respiratory arrest within minutes of envenomation. However, octopus's own Na$^+$ channels are resistant to TTX and even the females inject this neurotoxin into eggs to make them generate their own venom before hatching.

For human beings, often the venom proves fatal, to the extent of killing the victim within minutes. Initially, a bite from the octopus may be painless but severe pain, itchy sensation and urticarial lesions develop within a few minutes. Erythema and edema affect the extremities. Respiratory paralysis often leads to death.

Treatment is supportive, with pressure immobilization to reduce venom spread and intubation, if necessary. Because the venom primarily kills through paralysis, victims can be saved through artificial respiration or using a ventilator. Anybody who survives the first 24 hours recovers faster.

- **CUTTLEFISH:**

Inhabiting the sandy and muddy regions of shallow coastal waters from southern Great Barrier Reef to central South Australia, a *Cuttle fish (*Sepioloidea lineolata*), commonly known as **'Striped pyjama *Squid'** or the **'Striped dumpling squid'** (Fig. 3.2 e) is regarded both venomous and poisonous. They, respectively, get the common names from vibrant black stripes on its body, similar to **striped 'pyjamas'** (or Beetlejuice's suit) and appearing like a rounded **dumpling** (a small mass of soft dough that is boiled, fried or steamed; like momos). Not to be emphasized that they are small-bodied (7.0 to 8.0 cm) and rounded like a dumpling. On account of diagnostic colouration, this cuttlefish is a master of disguise and during daylight lay buried in sand mixed with broken shells so that only the top of its head and yellow eyes are visible. At night, it emerges to feed on small shrimp and fish. Their stripes are often used to warn predators away.

Whenever it is likely to be attacked by a predator, the mucus glands present on the skin ventrally, secrete a **toxic slime** (a mixture of proteins and toxins); poisonous enough not only to scare off the predator but to have a safe escape, also. Its venomosity is due to the **tetrodotoxin-like neurotoxic venom** present in the saliva.

* Both <u>Cuttlefishes</u> and <u>Squids</u> are ten-armed Cephalopds (Decapodiformes). **'True Cuttlefishes'** belong to Order Sepiida whereas 'Squids' belong to Teuthoidea (or Teuthida). The common name **'Pyjama Squid'** is is a surprising common name under 'cuttlefish' group. In reality it is a 'Cuttlefish', closely resembling with related cuttles, called **'Bobtail Squids'** (*Euprymna* sp.) from Indo-Pacific Ocean, Australia; belonging to another Order Sepiolida. 'Sepiolid' cephalopods appear to be 'all head' (= body rounded; therefore, known as dumpling squid or stubby squid). Though Sepiolids do not have cuttlebones (like true cuttles) but they are far more similar to cuttlefish than to other squids. **Striped Pyjama Squid** (*Sepioloidea lineolata*) is also a member of bobtail squids and because it also does not have a cuttle bone, it is named differently as 'squids'. Conclusively, they are not quite a cuttlefish (without a cuttlebone), but are not a squids either. Since, bobtail squids lack a 'true cuttlebone' and are without or with a reduced chitinous gladius (pen); they are regarded by some as the **sister group** of the **true squids**. Scientists once believed that cuttlefish were a completely separate lineage from other ten-armed cephalopods, however, recent genetic studies show that cuttlefish are evolutionarily among the groups of squids.

Chapter – 4
POISONOUS AND VENOMOUS FISHES
(XIPHOSURA AND CRUSTACEA)
Red swamp crayfish, *Procambarus clarkii, a watercdolour illustration*

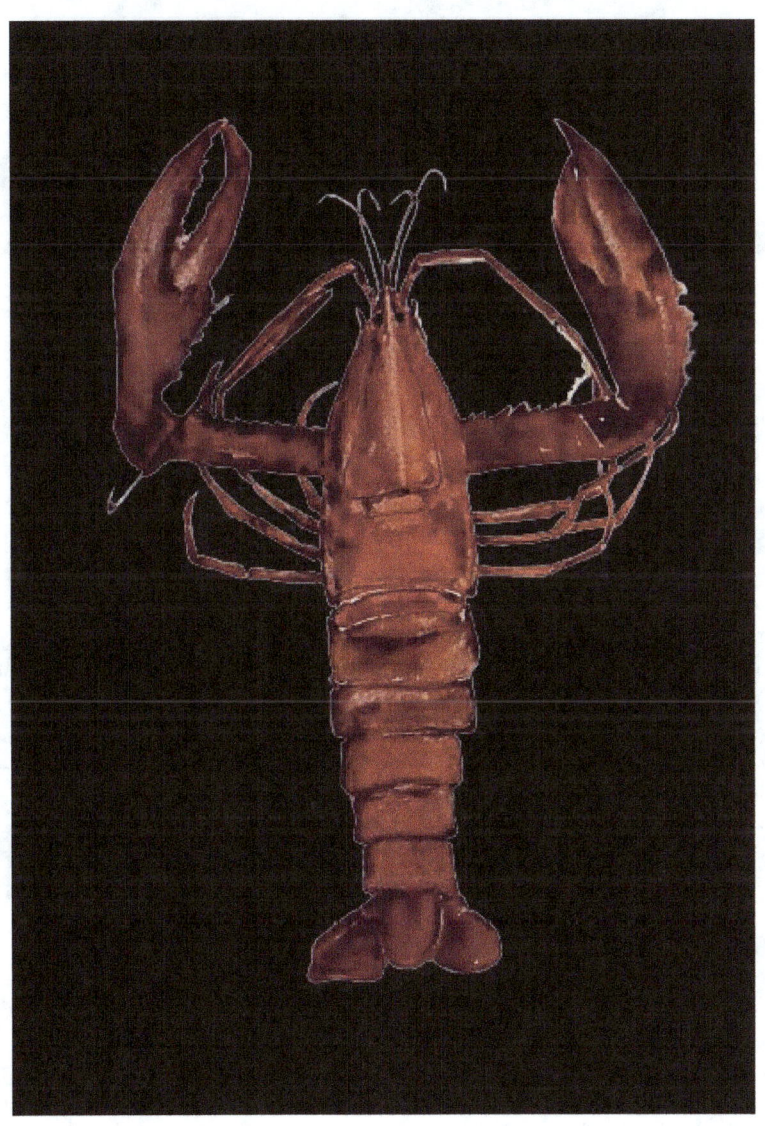

CHAPTER 4

POISONOUS AND VENOMOUS SHELLFISHES

[Xiphosura and Crustacea]

> **HIGHLIGHTS**
>
> XIPHISAURAN or HORSE-SHOE CRAB POISONING
> CRUSTACEAN SHELLFISH-POISONING
> Shrimps, Crayfish, Lobsters, Crabs, Prawns.
> VENOMOUS CRUSTACEAN

Besides Molluscan 'shellfishes', a good number of aquatic Arthropods classified as **Xiphosurans** (King Crabs) and **Crustaceans** (Crabs, Shrimps, Crayfish, Lobsters) are not only edible in most parts of the world but are **poisonous** or **venomous** from the 'defensive' strategy or human health hazards point of view.

4.1 XIPHISAURAN or HORSE-SHOE CRAB POISONING:

Any of various marine arthropods, belonging to **Subphylum – Celicerata, Class - Merostomata, Order – Xiphosura** and **Family - Limulidae**, are called **horse-shoe crabs**. They are more closely related to **Arachnids** (a group that includes spiders and scorpions) than to crustaceans (a group that includes true crabs, lobsters and shrimps). Horse-shoe crabs are aptly called **'living fossils'** because fossils of their ancestors date back to almost 450 million years and they are the only living representatives. Despite inhabiting the planet for so long, horse-shoe crabs changed very little over time. All of them live at sandy and muddy coastal stretches and during reproductive period they tend to migrate to the intertidal zone for egg-laying.

Fig. 4.1: Horseshoe Crab and the greenish egg mass (Roe).

Their body is divided into the anterior largest part, the **prosoma** or so-called 'head', the middle one triangular abdomen or **opisthosoma** (with spines on the sides) and long **sword-like tail** or telson (hence, **Xiphisaura**, < Gk. *xiphos* = sword + *oura* = tail) [Fig. 4.1]. The common name **'horse-shoe crab'** originates from the rounded **U-shaped head** (Prosoma), just like the shoe on a horse's foot. Other common names given to these crabs are the **horse-foot crab**, the **helmet crab**, the **saucepan crab**, as well as erroneously as the **king crab** (*Paralithodes camtschatica* is the true king crab).

Best known single **Atlantic/American species** is *Limulus polyphemus* (= *Xiphosura polyphemus*) attaining a length of more than 60.0 cm. The other three species, closely resembling *Limulus* in both structure and habits, are the **Chinese/Japanese** or **tri-spine horseshoe crab** (*Tachypleus tridentatus*), the **Indo-Pacific/coastal horseshoe crab** (*T. gigas*) and the **mangrove horseshoe crab** (*Carcinoscorpius rotundicauda*) found in Asia, from Japan to India.

In Malaysia the Horse-shoe Crab is a delicacy and there are specialist restaurants where it is the main item on the menu. The orange to green and tasting marinish **'Roe'** (eggs) is considered as an **aphrodisiac** in parts of Asia. The eggs of the Atlantic species are mainly **green in colour** (Fig. 4.1) and it is generally suggested that the green eggs be avoided. There are several ways to prepare dishes out of the **'Roe'**. The whole Horse-shoe Crab is grilled, laid on its back, the underside of the shell is peeled off from the front and the **roe** is extracted with a spoon and then consumed [https://www.colourbox.com/image/horseshoe-crab-eggs-image-9439384].

4.1.1 The poisoning:

Flesh, unlaid green eggs ('**Roe**') and viscera of these crabs become toxic during reproductive period. There have been a number of poisoning cases after ingesting Horse-shoe Crab **Roe** and even deaths have been reported from Thailand and Campuchia. Since 1994, horse-shoe crab-poisoning occurred both sporadically and epidemically in Chon Buri (Thailand). Similar cases of poisoning are reported from China. The allergic reactions have also been noticed after consuming **tri-spine horseshoe crab** (*T. tridentatus*) in China. This is attributed to a powerful **neurotoxin**, the **tetrodotoxin** (TTX), very similar to that found in **puffers** or ***Fugu*** (pl. refer to 'tetraodotoxin-poisoning' in Chapter – 1). As is well-known, TTX is produced by a number of marine bacteria spp. *viz.*, *Pseudomonas*, *Vibrio*, *Alteromonas*, *Shewanella* etc. and it is supposed that the TTX is accumulated *via* the food chain.

4.1.2 Symptoms:

In order of severity, the main symptoms are circumoral and lingual numbness, numbness of hands and feet, weakness, dizziness and vertigo, nausea and vomiting, transient hypertension, respiratory paralysis, fixed dialated pupils, ophthalmoplegia, blood pressure lower than 90/60 *mm* Hg and polyuria. The **allergic reactions** are similar to the reactions as those after consuming shrimp or crab (pl. see Chapter - 5).

4.1.3 Treatment:

Decontamination, symptomatic and supportive therapy is the usual remedy. The treatment of the poisoning from eating horse-shoe crab is similar to the treatment of TTX-poisoning by removing the toxins from body. Endotracheal intubation and mechanical ventilation are considered when paralysis progresses rapidly. The allergic reactions can be cured with antihistaminic **chlorpheniramine**.

4.1.4 Prophylaxis:

Eating horse-shoe crabs should be avoided during reproductive season.

4.2 CRUSTACEAN SHELLFISH-POISONING:

4.2.1 Shrimps:

Traditionally, decapod Crustaceans are divided into two suborders, the **Natantia** (swimmers) and **Reptantia** (walkers). **Shrimps** are included under **Natantia**. Some identify shrimp with the Infraorder - **Caridea**. More than 3000 Caridean species are '**true shrimps**', only about 20 of them being commercially important. In appearance, shrimps are very similar to prawns. They are found near the seafloor of most sea-coasts and estuaries as well as in rivers and lakes. Marine species are found at depths of up to 5,000 metres and from the tropics to the polar regions. Most shrimps are **omnivorous**, some are filter-feeders and others scrap algae from rocks. The others, called **cleaner shrimps**, feed on the parasites and necrotic tissue of the reef fish. On the other hand, they play a very important role in the food chain as being food sources for larger animals from fish to whales. Commercial shrimp species support an industry worth billion dollars a year. **Important commercial** spp. are *Penaeus monodon* (Tiger shrimp), *P. indicus* (Indian or white shrimp), *P. merguiensis* (Banana shrimp), *P. semisulcatus* (The green tiger or bear shrimp), *P. pencillatus* (Red-tailed shrimp), *P. japonicus*

(Japanese tiger shrimp), *P. vannamei* (Pacific white leg shrimp), *Metapenaeus monoceros* (Speckled shrimp) etc (Fig. 4.2). Their muscular tails are a delicacy and are, thus, widely caught and farmed for human consumption, being a rich source of *omega-3 fatty acids*. They are sold frozen, based on their categorization of presentation, grading, colour and uniformity.

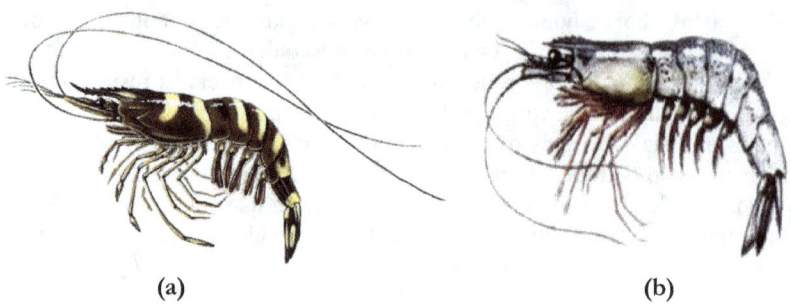

(a)   (b)

Fig. 4.2: Shrimps: **(a)** *Penaeus monodon*; **(b)** *Metapenaeus monoceros*.

(a) The poisoning:

Shrimps that are infected with bacteria, parasites, toxins or viruses are known to cause food-poisoning. People, who are always at a greater risk of food-poisoning, are the elders, infants; those who have a pre-existing medical history; those taking antibiotics, antihistamines or steroids; those with a weakened immune system and pregnant women. Some of the common causes of poisoning include **stale shrimp** (that are prepared even after the expiry date), **tinned** or **packed shrimps**, **diseased shrimps** (those harvested despite infections) or **chemicals in the shrimp's body** (due to use of chemicals in aquculture farms). More importantly, shrimps contain **Arsenic Pentoxide** (As_2O_5) and when any body takes **Vitamin C** after ingesting shrimp, the Arsenic Pentoxide present in the shrimp turns into **Arsenic Trioxide** (As_2O_3), proving to be highly poisonous. Resultantly, there is heart, liver, kidney and blood vessels' failure, leading to death.

(b) Symptoms:

Symptoms of poisoning become evident half an hour to several days after the consumption of shrimp. Stomachache, headache, vomiting, dizziness, weakness in the muscles; numbness in arms, legs, lips and tongue, blood in the stool, blurred or double vision, difficulty in swallowing etc. are the common symptoms.

(c) Treatment:

The common treatment strategies include *viz.*, fluid therapy *i.e.*, patients are needed to be hydrated (orally or itravenously) especially if there is vomiting or severe diarrhea; anti-emetic and anti-diarrheal medication and pain-releif medication. Light food for few days is suggested and the most appropriate diet is **BRAT** (**B**anana, **R**ice, **A**pple sauce and **T**oast).

(d) Prophylaxis:

Prior to buying and consuming the shrimp, expiry date should be checked and it should not be stored for a longer duration in refrigerator. Uncooked shrimp be avoided.

4.2.2 Crayfish:

The **red swamp crayfish**, **Louisiana crayfish** or **mud buga freshwater crayfish** (*Procambarus clarkii*), a native to the Southeastern United States, is now worldwide in distribution (Fig. 4.3). Tons of crayfish are consumed each year. Crayfish farming has become quite popular in China and with the increasing yield, a number of **crustacean-poisoning** cases have been reported *e.g.*, in July-August, 2010, some people from Nanjing (China) developed symptoms of **Haff disease/rhabdomyolysis** after

Fig. 4.3: Red swamp crayfish (*Procambarus clarkii*).

consuming this crayfish. **'Rhabdomylosis'** is characterized by disintegration of striated muscle fibers (a serious myopathy) and releasing the intracellular contents into the circulatory system; due to the toxicity of **Palytoxin** found accumulated in crayfish flesh. (for details pl. refer to Chapter 1).

4.2.3 Lobsters:

The term '**lobster**' generally refers to the clawed, large, marine crustaceans belonging to **Nephropidae** (including northern hemisphere *Nephrops* sp. and the southern hemisphere *Metanephrops* sp.) and **Homaridae** (including *Homarus* sp. from the northern Atlantic Ocean) [Fig. 4.4]. They have long bodies with muscular tails and live in crevices or burrows on the sea floor. Highly prized as seafood, lobsters are one of the most profitable commodities in coastal areas. Many lobsters, sold commercially, are killed and frozen before cooking. If the lobster is 'beheaded' before or soon after death, the flesh remains fresh for quite long because of the fact that the so-called head includes the thorax, the site of most of the viscera and gills, which is spoiled more rapidly than claw or tail meat.

(a) The poisoning:

Lobsters do not become poisonous if they die before cooking. Lobster's flesh needs to be cooked until it is opaque and milky-white. As lobsters are filter-feeders, they derive toxins, produced by marine algae. High levels of **paralytic shellfish poisoning** (PSP) toxins (**saxitoxin** and **gonyautoxin**, as already described with reference to molluscs) have been detected in the '**tomalley**' (< Carib. *tumale* = a sauce of lobsters' liver) or **lobster paste**, which is actually a soft **green gland** found in the body cavity, having combined functions of intestine, liver and pancreas (= **Hepatopancreas**). The term '**lobster paste**' is also often used to indicate a mixture of **tomalley** and **lobster roe**.Tomalley is considered a delicacy and is consumed alone but often added to sauces for flavour and as a thickening agent. The toxins responsible for most lobster poisonings are water-insoluble, heat and acid-stable and thus are not wiped off even after cooking/boiling. Time and again, it has been warned not to use lobster '**tomalley**', because it accumulates high levels of toxins and other pollutants. High levels of **Polychlorinated biphenyls** (PCBs) have also been found in tomalley, causing health problems when found in large concentrations.

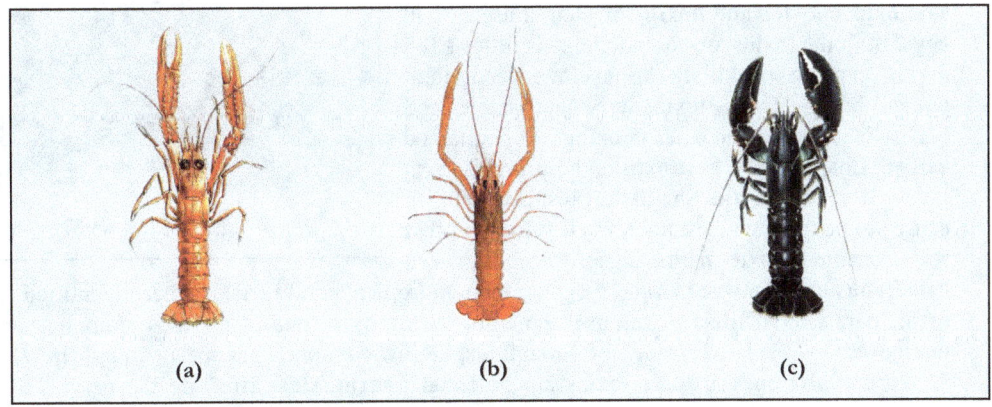

Fig. 4.4: Lobsters: (a) *Nephrops* sp.; (b) *Metanephrops* sp.; (c) *Homarus* sp.

(b) Symptoms:

The chief symptoms include a feeling of paralysis as well as difficulty in breathing, often leading to death if enough of the toxins are ingested. Vomiting, nausea, diarrhea and even permanent or short-term memory loss are the other common symptoms. (pl. also refer to PSP in Chapter- 1).

(c) Treatment:
Symptomatic

4.2.4 Crabs:

Under **Class – Crustacea**, **Order – Decapoda**, the **'true crabs'** belong to **Infraorder – Brachyura** (short-tailed) but the hermit crabs, king crabs and coconut crabs are classified under the **Infraorder – Anomura** (differently-tailed). Brachyuran crab-poisoning is chiefly caused by Xanthid crabs (Xanthidae), Blue Crabs (Portunidae) and Dungeness crabs (Cancridae) whereas the the most common poisonous hermit crabs, belonging to Anomura, are exemplified by Coconut Crabs (Coenobitidae) [Fig. 4.5 - 4.8].

There is a common saying that '**a dead crab is a bad crab**'. When crabs die, they degenerate so quickly, as to release toxins into all parts of body. If such crabs are in a tank/pond, the toxins kill other crabs as well, if the dead ones are not removed immediately. When any one ingests a dead (toxic) crab he/she is bound to be affected by toxins. Food-poisoning is due to ingesting bad or under-cooked crab meat.

The poisoning from crabs can be avoided by adopting various measures *viz.*, while purchasing, it must be ensured that the crab is lively; slow sluggish crabs should be avoided; if the crabs are to be transported for longer duration/distances they are to be kept cool and wet; water should be sprinkled over them at regular intervals, but need not be submerged and be kept away from sunlight; they must be cooked immediately or killed, cleaned and put in deep freezer to be cooked within a day.

- **Xanthid Crabs** (Fig. 4.5):

Belonging to **Family – Xanthidae,** most of the **toxic crabs** are found distributed in Indo-Pacific region. They includes the **most colourful**, shore-dwelling large crabs, the black tipped claws being diagnostic to them. The important sp. are *Atergatis floridus, Carpilius convexus, C. maculatus, Demania alcalai, D. toxica, D. splendida, Eriphia sebana, Lophozozymus pictor, Platypodia granulosa, Zosimus aeneus, Ranina ranin,. Daldorfia horrida, Thalamita prymna, T. danae, Pilumnus vespertilio, Ocypode ceratophthalma* etc.

(a) **The poisoning:**

Zosimus aeneus (Toxic Reef Crab or Devil Crab), *Platypodia granulosa* and *Atergatis floridus* are known to be **deadly poisonous**. They accumulate two of the most lethal natural toxins, the **saxitoxin** and **tetraodotoxin**, in their muscles and eggs. Both the toxins are so toxic, that as little as 0.5 *mg* is capable of killing an average-sized adult human being. These toxins are heat stable and persist in tissues even after cooking. As explained earlier (Chapter – 1), **saxitoxin** is the primary toxin involved in **Paralytic Shellfish Poisoning** (PSP), caused to people by ingesting mussels or oysters that have consumed toxic marine algae. Recently, a **red calcareous alga/seaweed** (*Jania* sp.), common on tropical coral reefs, has been identified as the source of the paralytic shellfish toxins in these crabs. **Palytoxin** (one of the most complicated and potent marine toxins isolated from soft zoanthids, *Palythoa* spp.) has been found in *Demania* and *Lophozozymus* spp. All these toxins are potential **neurotoxins**, affecting the nerve cells' ability to transmit information, thus resulting in paralysis.

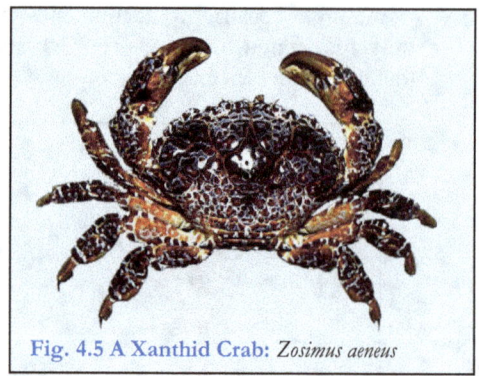

Fig. 4.5 A Xanthid Crab: *Zosimus aeneus*

(b) **Symptoms:**

Poisoning symptoms are similar to those as described for **Paralytic Shellfish Poisoning** (PSP) by molluscan shellfish poisoning. Numbness of the tongue and vomiting are the symptoms to appear first. Numbness of limbs gradually progresses to paralysis. Death occurs as a result of respiratory failure, usually within 12 hours. Mortality rate has been found to be over 50%.

On the other hand, persons intoxicated by ***Lophozozymus pictor*** suffer from vomiting, dizziness, abdominal and muscle pain. Numbness of the lower extremities, back pain, spasms on the face, hands and extremities appear 30 min. before death. The **fatality rate** has been assessed to be 80%.

Incidence of fatality resulting from the ingestion of **Demania toxica** has been reported in Philippines. The victims developed severe diarrhea, nausea and vomiting, accompanied with hypersalivation, frequent expectoration and muscular exhaustion. Finding it increasingly difficult to speak or breathe, convulsions developed and respiration ceased, ultimately resulting into death.

- **Blue Crabs** (Fig. 4.6):

Callinectes sapidus (*Gk. calli-* = beautiful, *nectos* = swimming, and *L. sapidus* = savory *i.e.*, pleasant or agreeable in taste or smell), the **Chesapeake blue crab** or **Atlantic blue crab** or simply **blue crab** (23.0 cm), is native to the Atlantic Ocean from Nova Scotia to Argentina and around the entire coast of the Gulf of Mexico. It has been introduced into Japanese and European waters and has been found up to the Baltic, North Mediterranean and Black Seas. The **blue hue** to the exoskeleton is imparted by the pigment called **alpha-crustacyanin** present in the shell, and interacting with a red pigment, **astaxanthin**, to impart a greenish-blue colouration. When cooked, the alpha-crustacyanin breaks down, leaving only the astaxanthin and turning the crab into red-orange or hot pink.

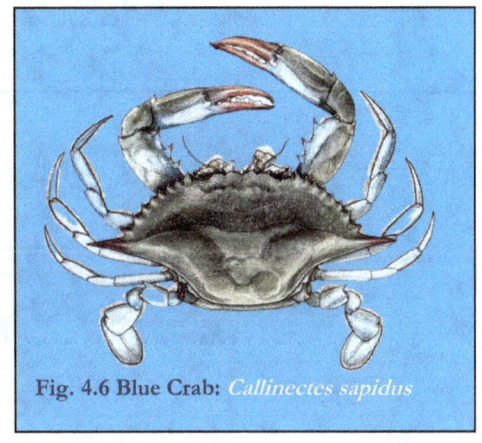

Fig. 4.6 Blue Crab: *Callinectes sapidus*

It is an omnivore, typically consuming thin-shelled bivalves, annelids, small fish and plants, including carrion (rotten flesh), individuals of its own population (**cannibalistic**) and animal waste. It undertakes seasonal migration for mating.

(a) **The poisoning:**

Consumption of **green gland** or **'tomalley'** often becomes hazardous. As described for lobsters, the **tomalley** (= hepatopancreas) of crabs is yellow or yellow-green in colour. In Maryland and on the Delmarva Peninsula (US), the hepatopancreas of the **blue crab** is called the '**muster**' or '**mustard**', because of its yellow colour. The tomalley is consumed mostly after moderation, particularly when the crabs are steamed or boiled. However, it is usually found to contain high levels of **polychlorinated biphenyls** (PCBs), responsible for a number of health problems. The **pollutants** such as **PCBs**, **dioxin**, **mercury** etc. accumulate in the crab's fatty tissues and concentrate in the hepatopancreas. Other toxins like **saxitoxin** and **gonyautoxin**, found associated with **Paralytic Shellfish Poisoning** (PSP) have also been reported in blue crab-poisoning incidents. These toxins do not leach even after cooking the crabs in boiling water.

The **bitter crab disease** (BCD) is on account of **'bitter flavour'** after boiling the crabs parasitized by a dinoflagellate **blood parasite**, *Hematodinium perezi* (BOX 4.1).

(b) **Symptoms:**

Same as mentioned with PSP and Lobster-poisoning *viz.*, nausea, a sense of floating and tingling sensations on the lips, tongue, fingers and toes.

(c) **Treatment:**

Symptomatic.

- **Dungeness crabs** (Fig. 4.7):

Metacarcinus magister (upto 25 cm), the **Dungeness crab** (named after the port of Dungeness, Washington), inhabits eelgrass beds on the west coast of North America and is the **largest edible species** (hence, known as **Market Crab** or **Common Edible Crab**) from Alaska to California, making this species important for commercial fisheries. The commercial Dungeness crab fishery began around 1916 in Alaska. Today, they are canned, frozen, shipped fresh/live to the market. A **Dungeness Crab** and **Seafood Festival** is held in

Fig. 4.7 Dungeness Crab: *Metacarcinus magister*

Port Angeles every year in October. About 1/4th of the crab's weight is sweetish, tender meat. Its diagnostic **red-brown** to **purple carapace**, with a spine-tipped edge on the anterior half, contains 10 teeth along the anterolateral margins. The chelipeds are purple to brownish at the base and tipped white. Dungeness crabs are sold either live or cooked. Live crabs are cooked simply by dropping them into boiling salt water. It is carnivorous and feeds chiefly on small clams, oysters, fish, shrimp, worms etc.

> **BOX 4.1**
> **BITTER CRAB DISEASE (BCD)**
>
> The disease gets its name on account of the fact that when cooked, the **blue crab** shows a chalky texture and a **bitter aspirin-like flavour**, compelling the consumer to discard it. Except the bad taste, this disease has no known impact on its consumers. Small number of diseased crabs can render an entire batch of crabs unpalatable. **BCD** or **Bitter Crab Syndrome** is caused by a dinoflagellate blood parasite, *Hematodinium perezi*. Once infected, the parasite grows rapidly inside the crab (up to 100 million parasites/*ml* of blood) over the course of 3 to 6 weeks. The **crab's blood changes to milky-white and loses it clotting ability**. Also, the infected crabs have drooping limbs and mouthparts.
>
> The parasite consumes oxygen from the crab's blood and tissues, rendering it to become weak and lethargic and to eventually die. *Hematodinium* infections are, therefore, highly pathogenic. In naturally and experimentally infected blue crabs, mortality has been noticed up to 87% over 40 days.
>
> Severe economic losses have been attributed to this parasite. Therefore, proper disposal of infected crabs is essential in controlling dissemination of the parasite. Control strategies include culling on station or within a watershed, culling or removing dead animals to onshore fertilizer processing plants, limiting transportation of live animals and not using potentially infected crabs for bait.

(a) **The poisoning, Symptoms etc.:**

Most people consume un-eviscerated Dungeness crab contaminated with **Domoic Acid** (DA), which causes **Amnesic Shellfish Poisoning** (ASP) in people (as also found in case of mussel and clam poisoning). DA produced by an algae (*Pseudo-nitzchia*) accumulates primarily in the viscera or **'crab butter'** or **'tomally'** (hepatopancreas).

Besides ASP, **Paralytic Shellfish Poisoning** (PSP) has also been of common occurrence in these crabs from the commercial fishery around Kodiak Island and the southern Alaska Peninsula. In 2010, for the first time, fatality occurred due to PSP to an Alaska resident after consumption of dungeness crab viscera. Closely related to blue mussels, the **Bay mussel** or **Foolish mussel** (*Mytilus trossulus*) was found to be the contaminant food source of these crabs (due to accumulation of the **saxitioxins** produced by dinoflagellates like *Alexandrium*, *Pyrodinium* and *Gymnodinium* sp.).

In initial stages of ASP, the vicitim experiences intestinal distress. Severe ASP can cause a facial grimace or chewing motion, short-term memory loss and difficulty in breathing.

Symptoms of PSP initially involve numbness and a burning or tingling sensation of the lips and tongue that spreads to the face and fingertips. This leads to a general lack of muscle coordination in the arms, legs and neck. Severe cases of PSP have resulted in respiratory paralysis and death.

There is no cure and treatment is strictly to provide comfort to the patient.

Fig. 4.8: The Coconut crab: *Birgus latro* on a coconut tree.

NOTE: For detals pl. refer to ASP and PSP poisonings in Chapter – 3.

- **Coconut crabs** (Fig. 4.8):

The **Coconut crab** (*Birgus latro*), commonly called **terrestrial hermit crab** or **robber crab** or **palm thief,** is the **largest terrestrial arthropod** (1.0 *m* from leg to leg, weight of up to 4.1 *kg*), found on islands across the Indian Ocean and parts of the Pacific Ocean as far east as the Gambier Islands. They enter water only for reproductive purposes and usually feed on plants, including coconuts that they cut down from the tree themselves. It is considered a **delicacy** and an **aphrodisiac** by Southeast Asians and Pacific Islanders and thus, extensively hunted.

The crab is rendered **toxic** after feeding on deadly poisonous leaves, fruits and seeds of an evergreen coastal tree, called *<u>sea mango</u> (*Cerbera manghas*; growing up 12 *m*). The said plant is the source of a toxic cardiac-arresting steroid called **cardenolide glycoside**.

Depending on the concentration of poison ingested, the same symptoms appear as seen in **paralytic shellfish poisoning** (saxitoxin) or **tetrodotoxin** poisoning.

> *In old days, people were to use the sap of the tree as a poison for animal hunting. The fruit was eaten to commit suicide in the Marquesas Islands (French Polynesia). In Madagascar, the seeds were used in sentence rituals to poison kings and queens.

4.2.5 Prawns:

Usually, Prawns are not poisonous but often people may suffer from severe **allergies**. However, it is suggested not to eat prawn followed by taking **Vitamin 'C'**. Like shrimps, the **Arsenic Pentoxide** (As_2O_5) present in the prawns reacts with Vitamin C to produce **Arsenic Trioxide** (As_2O_3), which is highly poisonous. At the University of Chicago (US), the investigations have revealed that their carapace contains much higher concentration of **potassium arsenic compounds**. This kind of arsenic poisoning can cause paralysis of small blood vessels. Therefore, a person who dies of arsenic poisoning shows signs of bleeding from the ears, nose, mouth and eyes. Thus as a precautionary measure, eating prawn is forbidden while taking vitamin C.

4.3 VENOMOUS CRUSTACEAN:

Among Arthropods, venomous individuals are quite common in a number of species of insects, scorpions, spiders and centipedes. However, from amongst the Xiphosaurans (represented by 4 species) and about 70,000 species of Crustaceans (water fleas, crabs, lobsters, barnacles and brine shrimp), none of them has been known to be venomous.

However, the **first venomous predatory crustacean**, *Speleonectes* (=*Xibalbanus*) *tulumensis,* was described in 1981. **Class - Remipedia** (L. *remipedes* = oar-footed) was erected under Crustacea to accommodate these marine (Australia, the Caribbean and the Atlantic), blind, albino, underwater cave-dwelling, very small (1.0 – 4.0 cm), elongated (superficially centipede-like), segmented (about 32 segments), hermaphroditic, crustacean-feeder, the only known **venomous crustaceans** (Fig. 4.9 a). They are slow moving with typical biramous swimming appendages present on each segment.

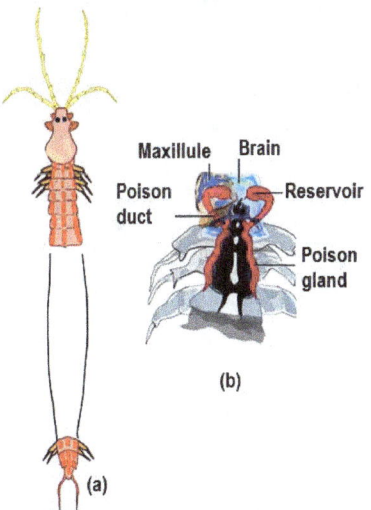

The **venom glands** are paired, located in the anterior trunk region, their ducts leading to reservoirs located in the terminal part of the maxillulae (Fig. 4.9 b). The contraction of the muscles attached along the anterior and posterior aspects of the reservoir creates a pressure to release the venom. The cocktail of **toxins** present in the venom is represented by **proteinases**, **chitinases** and a **paralytic neurotoxin**.

Fig. 4.9: (a) Venomous crustacean, *Speleonectes* (=*Xibalbanus*) *tulumensis*; (b) Head region [ventral view] enlarged, showing Venom apparatus.

In the process of prey capture and venom injection, the prey is first immobilized by the action of **neurotoxic**

peptides. Peptidases and chitinases break down the internal tissues and dissociate the muscles from the exoskeleton and enhance the paralytic effect by allowing the venom to spread further all through the body. The liquefied tissue is then sucked up by the remipede. This liquification process is often described to be similar to that found in a rattlesnake's venom.

CHAPTER 5

SEAFOOD ALLERGY

> **HIGHLIGHTS**
>
> **ALLERGIC SEAFOOD ITEMS & SUBSTANCES**
> Shellfish allergen, Fish allergen
> **SYMPTOMS**
> **DIAGNOSTICS**
> **TREATMENT**
> **PREVENTION**

Allergy is a kind of **hypersensitivity** or an overreaction of the immune system, often leading to **anaphylaxis** (pl. refer to BOX 1.1). These anaphylactic reactions are exhibited by a person to substances present in the environment and are usually harmless to most people. Such substances are known as **allergens** and are found in dust mites, pets, pollens, insects, ticks, moulds, some medications and even **'foods'**.

Seafood allergy is prevalent in those areas where seafood is an integral part of the diet. Allergenic reactions to seafood products are caused by **ingestion, handling** and **inhalation**. Seafood includes both **fish** *e.g.,* Salmon, Tuna, Mackerel, Sardines and Cod; and **shellfish** *e.g.,* Mussels, Clams, Octopus, Squids, Crabs, Shrimps and Lobsters, responsible for severe *Immunoglobulin E* (*IgE*)-mediated reactions, with **fish** being one of the top eight allergenic foods. Generally, *shellfish allergy is more frequent worldwide than fish allergy* and allergy to crustaceans is more common than allergy to molluscs. When a person with an allergy to a particular fish/shellfish is exposed, proteins in the fish/shellfish bind to specific *IgE* antibodies produced by the immune system. This triggers the person's immune defenses, leading to reaction and symptoms that can be mild or severe.

Per capita consumption of seafood, including fish and shellfish, is well above the global average for many Asian countries because of their coastal geographical location and rich seafood supply; the leading countries being Taiwan, Thailand, Singapore, Vietnam and Hong Kong. As per an estimate, Asian countries consume two-thirds of global seafood production.

5.1 ALLERGIC SEAFOOD ITEMS & SUBSTANCES:

Seafood is often consumed raw in Asian cuisines and this practice promotes sensitization to heat-sensitive seafood allergens which is not reported in western countries. Besides, Asians consume some seafood species that are not being consumed in other parts of the world. **Fish allergy** is less common in Asia than shellfish allergy, but its prevalence is still considerably high. **Shrimp** (*e.g., Penaeus monodon* and *Litopenaeus vannamei*) has been found to be the most commonly sensitized food allergen source in Hong Kong, China and India. As the consumption of fish and seafood has increased, so has the number of people found to be allergic to such food items *viz.,* fish fingers, nuggets; sushi and sashimi, salads, oyster sauce, fish sauce, including Worcester sauce (containing mostly anchovies); fish oils, prawn chips/crackers etc. Unlike many food allergies, shellfish allergy is more likely to develop in adults than in children. People who are allergic to one class of seafood can often tolerate those from another *e.g.,* most people allergic to crustacean shellfish are able to consume fish. Likewise, people allergic to tuna can consume prawns.

Allergic reactions may also result when a susceptible person is not consuming an allergenic substance but by exposure to vapours emanating from cooking seafood or even preparation or handling. This is because fish and seafood release **amines** (amine vapours) during cooking process causing allergic reactions in respiratory tract and lungs. About 15% of people with seafood allergy react to vapours and steam produced during cooking (especially grilling and on the barbeque).

- **Shellfish allergen:**

The filamentous muscle protein **Tropomyosin** was first identified as a **shrimp allergen** in 1993. Light chain Myosin, sarcoplasmic calcium-binding protein, Troponin 'C', Triosephosphate Isomerase and Fatty acid-binding proteins are other shellfish allergens, although their physiochemical properties remain largely unknown. **Arginine kinase** and **Glyceraldehyde 3-phosphate Dehydrogenase**, on the other hand, were identified as other shrimp allergens (*e.g.,* in *Macrobrachium rosenbergii*). Unlike shrimps, both Tropomyosin and Arginine Kinase are major allergens in crabs.

Paramyosin is another heat-labile protein allergen restricted to some Asian (Japanese and Koreans) populations where **abalones** are served as sashimi.

- **Fish allergen:**

Various fish species like **Cod**, **Salmon** and **Tuna** are consumed worldwide and many inland Asian countries rely mostly on freshwater aquaculture products like **Carps**, **Tilapia** and **Catfish**. Like Shrimps, **Tropomyosin** is also registered as a **fish allergen** in the WHO/International Union of Immunological Scocieties (IUIS) database. The allergenicity has only been demonstrated in **Mozambique tilapia** (*Oreochromis mossambicus*). On the other hand, a calcium-binding muscle protein '**Parvalbumin**' was first identified as the **major fish allergen**. Similar to Tropomyosin, parvalbumin has been found resistant to thermal treatment and enzymic hydrolysis. Despite high homology across fish species, parvalbumin had different allergenicity in different fish species and the allergenicity of individual fish species is also dependent on the quantity of parvalbumin present in the muscle. **Collagen** is another heat-stable fish allergen from muscle and skin of fishes. Two heat-labile muscle enzymes of fish *viz.,* **aldolase** and **enolase**, have recently been identified as allergens but their clinical relevance remains uncertain. Apart from muscle proteins, **fish roe** is also reported to cause allergic reactions due to the presence of vitellogenin as yolk protein.

5.2 SYMPTOMS:

Symptoms of allergy appear within minutes up to two hours after eating fish/shellfish and may include **skin reactions** such as **urticaria** (hives) or **eczematous contact dermatitis**, allergic conjunctivitis (itchy, red, watery eyes), gastro-intestinal tract reactions such as nausea, abdominal pain, vomiting or diarrhea; respiratory tract problems like **wheezing** (high-pitched, coarse whistling sound produced on breathing), coughing, runny nose; troubled breathing, anxiety, distress, faintness, paleness, weakness; lowering of blood pressure, fast heartbeat, loss of consciousness, **angioedema** (swelling of lips, tongue or face) etc. Of them, the **skin** and **respiratory problems** need a special mention.

Workers, who have a job of processing fish/shellfish, suffer a great deal from asthma, rhinitis and conjunctivitis. An **occupational disease** is most prevalent in fish and seafood-handlers. Among the caterers, skin reactions from contact with fish/shellfish are the most commonly reported reactions. Among **major skin allergies,** contact urticaria is associated with direct contact of raw seafood proteins. At least 75% of **eczematous dermatitis** in the fish-processing industry is of an irritant nature, due to contact with water and fish products (fish juice, slime, skin and fillet). Contact with the proteinacious seafood also causes a **chronic recurrent dermatitis** commonly known as **Protein Contact Dermatitis** (PCD). However, biochemical sensitizers (*e.g.,* garlic, onion, spices) added to seafood can also cause a delayed allergic contact dermatitis. Predominantly affected areas are the palm and dorsal face of the hands. In more severe cases, local skin contact with seafood may result in generalized urticaria and/or systemic symptoms like angioedema and wheezing. It has been estimated that caterers, who suffer from such skin diseases, have to find alternative employment because of the severity of their symptoms.

A higher prevalence of **occupational asthma** is associated with exposure to **aerosols** arising from arthropods (crab and shrimp) than to molluscs and bony fish. **Reactive Airways Dysfunction**

Syndrome (RADS) is an **asthma-like illness**, often induced by exposure to high concentration of irritants such as **sulfite preservatives** or **ammonia** used as a refrigerant in seafood industries.

5.3 DIAGNOSTICS:

Skin and blood tests are the best methods to confirm the symptoms and to rule out the suspective food poisoning. **Skin prick test** is done where the skin is pricked and exposed to small amount of the proteins found in seafood, that anybody is allergic to. If any inflammation or reaction is observed, the patient is diagnosed to be positive for seafood allergy. **Blood** is tested for the presence of **allergen-specific immunoglobulin** (*IgE*) antibodies to a specific protein from the fish/shellfish being tested.

5.4 TREATMENT:

Currently, there is no specific cure for seafood allergies. Treatment of an acute allergic reaction includes prescription of an **antihistamine**, which helps reducing skin rashes, swellings and hives. Appropriate emergency treatment of **anaphylaxis** includes injection of **epinephrine** (adrenaline), which is available as an **autoinjector device** (commonly called an *Epi-Pen*). Epinephrine must be administered as soon as symptoms of a severe allergic reaction appear. If an allergist has diagnosed a food allergy and prescribed Epinephrine, it is advised to be carried with the patient all the time.

5.5 PREVENTION:

The allergen which causes symptoms after consuming crustaceans is quite resistant to cooking and after washing may even persist on cutting boards etc. Patients, with a history of severe allergic reactions to crustaceans, should avoid locations where these are cooked and served as these sensitive patients may react to allergens present either in vapours from cooking or remaining on surfaces after cleaning.

Reading labels of over the counter products, including 'natural' remedies and herbal supplements is also important *e.g.*, some weight loss products contain small amounts of a shellfish derivative called *Chitosan (BOX 5.1). **Glucosamine**, commonly used in the products for **osteoarthritis**, is also derived from the shells of crabs and lobsters. While there is no evidence that Chitosan and Glucosamine are harmful to persons who are allergic to shellfish, it is advisable to avoid these products until proper advice by the experts. The **Food Allergy Labeling Law** (FALCPA) is required to be abide by, which defines that a crustacean shellfish is one of the allergens. This means that manufacturers are required to list the presence of clams, oysters, mussels, scallops or other molluscs in ingredient lists.

BOX 5.1
***CHITOSAN**

It is a **linear nitrogenous polysaccharide** composed of randomly distributed *β-(1-4)-linked D-glucosamine* (deacetylated unit) and *N-acetyl-Dglucosamine* (acetylated unit). It is produced commercially by **deacetylation of chitin**, the structural element in the exoskeleton of crustaceans (crabs, shrimp, lobsters etc.) and cell walls of fungi. **Chitosan** is used to treat obesity, high cholesterol and Crohn's disease. It is also used to treat complications that kidney failure patients on dialysis often face, including high cholesterol anemia, loss of strength and appetite and troubled sleeping (insomnia). Besides, it can also be used in agriculture as a seed treatment and biopesticide, helping plants to fight fungal infections. In wine-making it is used as a clearing agent, also helping to prevent spoilage. In industry, it can be used in a selfhealing polyurethane paint coating.

QUESTIONS

I. What is Ichthyotoxism? How many categories of fish poisonings are there? Give an account of Ichtyosarcotoxic fishes.
II. Give an account of Scombroid and Ttetraodotoxin-poisoning.
III. Write an essay on Ciguatera poisoning.
IV. What is shellfish poisoning? Discuss about Molluscan shelfish-poisoning.
V. Write an essay on Crustacean shellfish poisoning.
VI. Giving a phylogenetic basis, distribution and habitat preferences of Venomous fishes, discuss about Cartilaginous venomous fishes.
VII. Give an account of general structural organization of the venom apparatus, toxicity mechanisms among catfishes.
VIII. Giving distribution, habitat preferences, epidemiology of venomous fishes, in general, give an account of the diversity of spiny-rayed, bony venomous fishes.

IX. Short-answered questions:
1. Ichthyotoxism and Ichthyoacanthotoxism.
2. Diarrhoeic shellfish toxins causing diarrhoeic shellfish poisoning (DSP).
3. Ichthyocrinotoxic fishes Amnesic shellfish toxins causing amnesic shellfish poisoning (ASP).
4. Ichthyohepatotoxic fishes Neurotoxic shellfish toxins causing neurotoxic shellfish poisoning (NSP).
5. Ichthyootoxic fishes Azaspiracid shellfish toxins causing azaspiracid shellfish poisoning (AZP).
6. Clupeotoxic and Gempylotoxic fish
7. Horse-shoe crab poisoning
8. Ichthyoallyeinotoxic fish
9. Shrimp-poisoning
10. *Gambierdiscus toxicus*
11. Lobster-poisoning
12. Symptoms and treatment of Ciguatera
13. Ciguatera Shellfish Poisoning (CSP).
14. Saber-toothed Blennies
15. Weeverfishes
16. Scorpionfishes
17. Stargazers
18. Remipedia
19. Tetramine

X. Multiple-choice questions :
1. The publication entitled 'Poisonous and Venomous Marine Animals of the World' in three volumes is authored by:
(a) J.S. Nelson (b) Perriere and Perriere (c) J.R. Norman (d) Bruce Walter Halstead
2. Which of the following are mainly Ichthyohepatotoxic?:
(a) Cyprinids (b) Catfishes (c) Sharks (d) Cyclostomes
3. Which of the following are mainly Ichthyocrinotoxic?:
(a) Hagfishes (b) Rays (c) Ratfishes (d) Dogfishes
4. Pahutoxin has been identified in which of the following?
(a) *Octracion* (b) *Diodon* (c) *Tetraodon* (d) *Carcharodon*
5. Toxins first isolated from the sea bass or soapfish (*Pogonoperca punctata*) are named as:
(a) Grammistins (b) Saponins (c) Ciguatoxins (d) Pahutoxins
6. *Schizothorax* sp. are:
(a) Hepatotoxic (b) Crinotoxic (c) Ootoxic (d) Hemotoxic
7. Dinogunellin has been identified as a lysophospholipid toxin in the roe of :
(a) Japanese prickleback (b) Shark (c) Japanese sandfish (d) Japanese Mackerel
8. Which of the following families is supposed to be hemotoxic?
(a) Cyprnidae (b) Anguillidae (c) Tertaodontidae (d) Siluridae
9. The term anaphylaxis was coined by:
(a) Charles Darwin (b) Charles Robert Richet (c) Halstead (d) Robert Rover
10. Histamine poisoning is caused by which of the following?
(a) Scombridae (b) Anguillidae (c) Muraenidae (d) Acipenseridae

11. Balloonfish, blowfish, bubblefish, globefish, swellfish, toadfish, toadies, honey toads and sea squab are the common names for the fishes belonging to:
(a) Coryphaenidae (b) Xiphiidae (c) Diodontidae (d) Tetraodontidae

12. The skin and certain internal organs of many fishes belonging to Tetraodontidae are highly toxic but nevertheless the meat of some species is considered a delicacy in both Japan (as *fugu*) and Korea (as *bok*). These fishes belong to Genus:
(a) *Tetraodon* (b) *Diodon* (c) *Stichaeus* (d) *Scorpaenichthys*

13. Which of the following is the main cause of clupeotoxism?
(a) Palytoxin (b) Ciguatoxins (c) Maitotoxins (d) Grammistins

14. Gempylid Diarrhea or Escolar diarrhea is caused by :
(a) *Gempylus* sp. (b) *Ruvettus* sp. (c) *Rutilus* sp. (d) *Rohtee* sp.

15. The older name of Tokyo (Japan) is :
(a) Edo (b) Shimonoseki (c) Tokugawa (d) Kansai

16. Which of the following is most famous for consumption of *'fugu'* in Japan?
(a) Tokugawa (b) Kansai (c) Shimonoseki (d) Tokyo

17. The most dangerous part of the 'Puffer' is :
(a) Liver (b) Ovary (c) Pancreas (d) Muscles

18. Full form of CFP is:
(a) Clupeoid fish poisoning (b) Ciguatera fish poisoning (c) Crayfish poisoning (d) Congereel fish poisoning

19. *Fugu*, that has its poisonous parts removed, is marketed with the name:
(a) Sashimi (b) Higanfugu (c) Migaki fugu (d) Takefuku

20. The *fugu*, cooked in a broth made from the poisonous livers and intestines, is called:
(a) Fugu Kara-age (b) Hire-zake (c) Chiri (d) Yubiki

21. 'Dream fish' or 'nightmare fish' are concerned with :
(a) Ichthyootoxism (b) Ichthyohepatotoxism (c) Ichthyocrinotoxims (d) Ichthyoallyeinotoxism

22. After ingestion, which of the following cause 'psychoactive' effects?
(a) *Kyphosus* sp. (b) *Erilepis* sp.(c) *Stromuteus* sp. (d) *Ruvettus* sp.

23. Dimethyltryptamine (DMT), found in some fish, is potentially a:
(a) Ciguatoxin (b) Tetraodotoxin (c) Scaritoxin (d) Hallucinogen

24. The name ciguatera was given by :
(a) Don Antonio Parra of Japan (b) Don Parra Antonio of the Carribean
(c) Antonio Don Parra of Germany (d) Don Antonio Parra of Cuba

25. The *'cigua'*, is the Spanish trivial name of :
(a) A Shrimp (b) A Crab (c) Scorpionfish (d) A Mollusk

26. Which of the following families are considered key vectors in the transfer of ciguatoxins to carnivorous fish?
(a) Acanthuridae and Scaridae (b) Carangidae and Labridae (c) Lutjanidae and Scombridae (d) Serranidae and Sphyraenidae

27. The dinoflagellate, known as *Gambierdiscus toxicus*, is potentially involved with:
(a) Ichthyoallyeinotoxism (b) Puffer fish poisoning (c) Ciguatera (d) Moray eel poisoning

28. Gambiertoxin is oxidized to a more oxidized form of toxin called:
(a) Tetraodotoxin (b) Ciguatoxins (c) Scaritoxin (d) Palytoxin

29. The main ciguatoxins identified from the Pacific are P-CTX-1, P-CTX-2 and P-CTX- 3, but the major toxin found in carnivorous fish is/are:
(a) CTX-1 (b) P-CTX-2 (c) P-CTX-3 (d) CTX-1 and P-CTX-2

30. Ciguatoxins solely affect the:
(a) Blood (b) Muscles (c) Peripheral nervous system (d) Respiratory system

31. Select out the fish, called *'maito'* in Tahiti language, after which ciguatera-causing maitotoxins (MTXs) have been named:
(a) *Ctenochaetus striatus* (b) *Calamus calamus* (c) *Caranx latus* (d) *Cheilinus undulatus*

32. Ciguatera-causing palytoxin was first isolated from:
(a) Diatoms (b) Dinoflagellates (c) Soft coral (d) Mollusks

33. Haff disease or Rhabdomylosis is caused by the poisoning from:
(a) *Tetraodon* (b) *Ictalurus* (c) *Ictiobus* (d) *Anguilla*

33. The major symptoms of Rhabdomylosis are associated with :

(a) Heart (b) Skeletal muscles (c) Gastrointestinal tract (d) Liver
35. Phycotoxins are toxic compounds that enter into the food chain as components of :
(a) Small fish (b) Soft corals (c) Mollusks (d) Phytoplankton
36. Saxitoxin (STX) is a neurotoxin. The term STX originates from which of the following species of bivalve mollusks?
(a) Alaska butter clam (b) Blue mussels (c) Soft-shell clams (d) Razor clams
37. More than 100 years ago, Paralytic Shellfish Poisoning (PSP) was, for the first time, reported in:
(a) China (b) Africa (c) South America (d) Canada
38. Which of the following are recognized as closely related 3,4,6-trialkyltetrahydropurine compounds?
(a) DSP toxins (b) PSP toxins (c) ASP toxins (d) NSP toxins
39. Which on of the following is listed as a grade one chemical weapon under the UN Chemical Weapons Convention?
(a) Saxitoxin (b) Tetrodotoxin (c) Okadaic acid (d) Pectenotoxin
40. Most of the effects of Saxitoxins (STXs) are on:
(a) Smooth muscles (b) Skeletal Muscles (c) Gastrointestinal tract (d) Peripheral nerves
41. Although DSP is reported worldwide, the highly affected areas appear to be:
(a) Japan and China (b) America and Africa (c) Asia and South America (d) Europe and Japan
42. DSP toxins are produced by dinoflagellates belonging to:
(a) *Pyrodinium* spp. (b) *Dinophysis* spp. (c) *Gymnodinium* spp. (d) *Alexandrium* spp.
43. Depending on chemical structure, DSP toxins are often divided into three groups. The first group includes:
(a) Polyether-lactones (b) Okadaic acid (c) Yessotoxins (d) Saxitoxins
44. Amnesic Shellfish Poisoning (ASP) is also called:
(a) Okadaic acid poisoning (OAP) (b) Azaspiracid poisoning (AZP)
(c) Domoic Acid Poisoning (DAP) (d) Dynophysistoxin poisoning (DTXP)
45. ASP was first recognised in 1987 in:
(a) Prince Edward Island, Canada (b) Bay of Fundy (c) Canada Oregon and Washington States (d) Japan
46. Domoic Acid (DA), the chief naturally occurring toxin responsible for Amnesic Shellfish Poisoning (ASP), is actually :
(a) A crystalline water-soluble alkaline amino acid (b) A crystalline fat-soluble acidic amino acid
(c) A crystalline water-soluble aromatic amino acid (d) A crystalline water-soluble acidic amino acid
47. The shellfish poisoning toxin, Domoic Acid (DA), was originally isolated from:
(a) Green macroalga (*Alsidium corallinum*) (b) Red macroalga (*Chondria armata*)
(c) Diatom (*Amphora coffaeformis*) (d) Dinoflagellate (*Dinophysis*)
48. DA intoxication has been aptly called as Amnesic Shellfish Poisoning (ASP), because of its impact on:
(a) Muscles (b) Memory (c) Blood (d) Liver
49. Neurologic Shellfish Poisoning (NSP) is also known as:
(a) Domoic Acid Poisoning (DAP) (b) Diarrhoeic Shellfish Poisoning (DSP)
(c) Brevetoxin (BTX) poisoning (d) Paralytic Shellfish Poisoning (PSP)
50. The body of which of the following is relatively fragile and is readily broken down due to wave action along beaches, thus releasing the toxins :
(a) *Karenia brevis* (b) *Gambierdiscus toxicus* (c) *Alexandrium* spp. (d) *Dinophysis* spp.
51. AZP toxins are found to have been produced by which of the following ?
(a) *Alsidium* sp. (b) *Dinophysis* sp. (c) *Protoceratum* sp. (d) *Gambierdiscus* sp.
52. Which of the following Horse-shoe crabs, causing shellfish-poisoning, is found along the Eastern coast of India?
(a) *Carcinoscorpius rotundicauda* (b) *Tachypleus tridentatus*
(c) *Limulus polyphemus* (d) *Xiphosura polyphemus*
53. The species of northern hemisphere Genus of 'lobsters' is:
(a) *Homarus* (b) *Penaeus* (c) *Limulus* (d) *Nephrops*
54. 'Tomalley' of the 'lobsters' and 'crabs' refers to:
(a) Toxin (b) Hepatopancreas (c) Tail muscles (d) Gonads
55. Bitter Crab Disease (BCD) is concerned with:
(a) Xanthid crabs (b) Blue Crabs (c) Dungeness crabs (d) Coconut Crabs
56. Alpha-crustacyanin is a pigment characteristic to the shell of a kind of crabs, imparting:

(a) Ortange hue (b) Green hue (c) Spotted black-brown (d) Blue hue

57. *Metacarcinus magister* is the scientific name of:
(a) Blue crab (b) Xanthid crab (c) Horse-shoe crab (d) Dungeness crab

58. Which of the following is the largest edible marine species of crabs?
(a) *Callinectes sapidus* (b) *Demania toxica* (c) *Metacarcinus magister* (d) *Lophozozymus pictor*

59. Which of the following is famous as 'Market Crab'?
(a) Dungeness crab (b) Coconut Crab (c) Blue Crab (d) Robber crab

60. ASP and PSP are the chief poisonings caused by ingesting which of the following crustaceans?
(a) Lobsters (b) Dungeness crab (c) Prawns (d) Shrimps

61. Which of the following is the largest landliving arthropod?
(a) Horse-shoe crab (b) Dungeness crab (c) Coconut crab (d) Xanthid crab

62. *Birgus latro* is the scientific name of:
(a) Coconut crab (b) Dungeness crab (c) Blue crab (d) King crab

63. Cardiac-arresting cardenolides or cardenolide glycosides are the chief toxins found in the flesh of :
(a) Xanthid crabs (b) Blue Mussels (c) Oysters (d) Coconut crabs

64. Which of the following is basically an shellfish allergy-causing protein?
(a) Myosin (b) Tropomyosin (c) Troponin (d) Actin

65. Most venomous fish are:
(a) Marine, bottom-dwellers (b) Marine, coast-dwellers (c) Estuarine (d) Coral reef dwellers

66. The Dogfish Sharks with venomous dorsal fin spines belong to which of the following Genera?
(a) *Scoliodon* (b) *Sphyrna* (c) *Squalus* (d) *Somniosus*

67. River stingrays belong to which of the following families?
(a) Hexatrygonidae (b) Plesiobatidae (c) Dasyatidae (d) Potamotrygonidae

68. The venom glands are restricted to the spaces between the serrations on the dorsal / pectoral fin spines, as opposed to being found along the length of the spines, in which of following families of catfishes?
(a) Ictaluridae (b) Mochokidae (c) Doradidae (d) Plotosidae

69. Both in *Ictalurus* (Channel catfish) and *Noturus* (Madtoms), there is:
(a) An inflicting pectoral spine (b) An inflicting preopercular spine (c) An inflicting cleithral spine (d) An inflicting Dorsal spine

70. Belonging to Family Plotosidae, several marine and estuarine eeltail catfishes inflict stings with their venomous:
(a) Pectoral fin spines (b) Dorsal and Pectoral fin spines (c) Dorsal and pelvic fin spines (d) Opercular spines.

71. The venomous apparatus, composed of two hollow dorsal spines and a hollow opercular spine, associated with venom glands, is characteristic to:
(a) Stargazers (b) Rabbitfishes (c) Weeverfishes (d) Toadfishes

72. The most common type of venom apparatus, in the form of the fin ray spines of dorsal, anal and pelvic fins, with grooves on both sides, is found in:
(a) Eeltail catfishes (b) Channel catfishes (c) Scorpionfishes (d) Saber-toothed blennies

73. The venom apparatus consists of the two cleithral spines, in which of the following venomous fishes?
(a) Weeverfishes (b) Lionfishes (c) Channel catfishes (d) Stargazers

74. The venom apparatus consisting of Toxic buccal glands associated with canine-like teeth on the dentary are characteristic to:
(a) Blenniidae (b) Trachinidae (c) Scorpaenidae (d) Uranoscopidae

75. Scalpel sharp blades, found on the caudal peduncle, are characteristic to:
(a) Rabbitfishes (b) Surgeonfishes (c) Sabre-toothed blennies (d) Weeverfishes

ANSWERS TO Q. No. X.

1. (d); 2. (c); 3. (a); 4. (a); 5. (a); 6. (c); 7. (a); 8. (b); 9. (b); 10. (a); 11. (d); 12. (d); 13. (a); 14. (b); 15. (a); 16. (c); 17. (a); 18. (b); 19. (c); 20. (c); 21. (d); 22. (a); 23. (d); 24. (d); 25. (d); 26. (a); 27. (c); 28. (b); 29. (a); 30. (c); 31. (a); 32. (c); 33. (c); 34. (b); 35. (d); 36. (a); 37. (a); 38. (b); 39. (a); 40. (d); 41. (d); 42. (b); 43. (b); 44. (c); 45. (a); 46. (d); 47. (b); 48. (b); 49. (c); 50. (a); 51. (c); 52. (a); 53. (a); 54. (b); 55. (b); 56. (b); 57. (d); 58. (c); 59. (a); 60. (b); 61. (c); 62. (a); 63. (b); 64. (b); 65. (a); 66. (c); 67. (d); 68. (c); 69. (a); 70. (b); 71. (c); 72. (c); 73. (d); 74. (a); 75. (b).

ANSWERS TO THE MCQs
'Fish & Fisheries Digest Part – 3'
[Light & Electricity]

1(d), 2 (c), 3 (b), 4 (c), 5 (d), 6 (d), 7(b), 8 (c), 9 (d), 10 (b), 11 (a), 12 (d), 13 (a), 14 (c), 15 (a), 16 (b), 17 (a), 18 (b), 19 (b), 20 (a) 21 (c), 22 (c), 23 (b), 24 (d), 25 (b), 26 (b), 27 (d), 28 (a), 29 (a), 30 (d), 31 (c), 32 (d), 33 (a), 34 (b), 35 (d).

APPENDIX – I

Harmful Algal Bloom(s) [HABs]

Among scientific community, mass proliferation activity of toxic phytoplankton (in Oceans) is popular as **Harmful Algal Bloom(s)** [HABs]. Non-toxic microalgae attaining high biomass through population explosion can also cause **HABs** by imparting discolouration to seawater (like red tides), anoxia or imparting high content mucilage that negatively affect the environment and human activities. In addition, some species of Cyanobacteria produce specific toxins, causing liver poisoning, dermatitis and even cancers.

Not to be emphasized that the *'Cultural Eutrophication'* from domestic, industrial and agricultural wastes are the chief contributors to HABs. Alternatively, introduction of harmful species, climatic variability and expanding aquaculture practices are also taken as possible causes of HAB trends, world over. As a first step towards a global HAB status assessment, the work entitled *'Global harmful algal bloom status reporting'*, has been published in February, 2021 (Special Issue of the Journal, *'Harmful Algae'*, Vol 102: *https://unesdoc.unesco.org › ark:*). It has been revealed that there are about 10,000 beneficial marine phytoplankton distributed in the world's oceans and about 200 of them are adjudged producing harmful toxins often leading not only to human health hazards but also aquaculture fish-kills, thus, interfering the recreational activities in coastal or inland waters and causing great economic losses. Whales, porpoises, other aquatic and terrestrial animals can also become victims when they accumulate toxins *via* contaminated water, plankton or fish.

A number of international programs have been launched from time to time to study and manage HABs and their links with environmental changes. The creation of a **Harmful Algal Bloom Panel** by the **Intergovernmental Oceanographic Commission** (IOC) of UNESCO has been one such initiative. The **International Atomic Energy Agency** (IAEA), **International Council for Exploration of the Sea** (ICES), **North Pacific Marine Science Organization** (PICES) and the **International Society for the Study of Harmful Algae** (ISSHA), are the active partners of Harmful Algal Bloom Panel.

The *Harmful Algal Event Database* (HAEDAT; *http://haedat.iode.org*) is the only existing open access meta database holding information about harmful algal events from across the globe. HAEDAT is a component of the **Harmful Algal Information system** (HAIS) within the **International Oceanographic Data and Information Exchange** (IODE) of the IOC with prime cooperation by ICES and PICES.

HAEDAT holds records from the ICES area (North Atlantic) since 1985 and from the PICES area (North Pacific) since 2000. IOC regional networks in South America, South Pacific, Asia and North Africa are in the process of contributing to the HAEDAT database.

To be more precise, a harmful algal event of deep socio-economic concern is identified as or belonging to at least one of the following types:
- **Water discolouration, mucilage scum or foams produced by non-toxic or toxic microalgae.**
- **Biotoxin accumulation in seafood above levels considered safe for human consumption.**

- **Harmful algae-related precautionary bans on shellfish or other invertebrate harvesting or closures of beaches to safeguard human health.**
- **Any event where humans, animals and other organisms are negatively affected by microalgae.**

Majority of the harmful events recorded in **HAEDAT** occurred between 1985 and 2016. Except for North Asia and the Pacific, annual records of every region began to be maintained between 1985 and 1987. **HAEDAT** datasets of 43 countries is combined in 13 regions of Ocean Biodiversity Information System (OBIS). Despite the split done for European regions, they remained by far the ones with the biggest numbers on **HAEDAT** events registered.

As of 10 December 2019 a total of 9,503 **HAEDAT** events had been entered from across the globe, comprising 48% seafood biotoxin, 43% high phytoplankton counts and/or water discolourations causing a socio-economic impact, 7% mass animal or plant mortalities and 2% others (including foam and mucilage production).

Among all events linked to **seafood toxin syndromes**, Paralytic Shellfish Toxins (PST) accounted for 35%, Diarrhoeic Shellfish Toxins (DST) 30%, Ciguatera Poisoning (CP) and marine and brackish water cyanobacterial toxins each 9%, Amnesic Shellfish Toxins (AST) 7%, and others 10% (including Neurotoxic Shellfish Toxins (NST), Azaspiracid Shellfish Toxins (AZT) and toxic aerosols).

ABOUT THE AUTHOR

It is more than half a century, since 1969 when 1950 born author (**S. K. Gupta / Shashi Kant Gupta**), enfaced with the esteemed readers, started practicing Zoology (Fish & Fisheries) as a Post-Graduate student at the *alma mater* DAV (PG) College, Dehradun (now in Uttarakhand, then Uttar Pradesh, India). This practice led to his Professional life as a Teacher in a Post-Graduate College, imparting higher education in District Banda, Bundelkhand, Uttar Pradesh (India). The prestigious Pt. Jawaharlal Nehru College was that stepping rung of his Teaching Career there, lasting for about 30 years *w.e.f* 1971 until 2000, when the services were got Transferred to the same *alma mater* place, DEHRADUN, but in the sister College of DAV, the DBS (PG) College, on account of unilateral transfer facility by the then UP Govt. The destiny destined serving the 'home place' during 2000 to 2010, a period of rendering services to the eliteclass-oriented families of DEHRADUN, a famous English Medium education hub. Definitely, this was a different experience as compared to teaching the poor & less educated-class of Bundelkhand region.

The author's earned experiences of long professional life are divisible into pre- & post-retirement phases. The pre-retirement phase was not only devoted to intense teaching but also to doing D.Phil. (1982), publishing research papers, writing articles in Hindi as well as English, carrying-out minor or major Projects etc. The post-retirement phase has also not been stationary / static / resting phase. To quench the thirst of presenting something before the academia, a septuagenarian 'Teacher' is still busy with doing more and more creative, constructive and rendering services to the 'Subject', Zoology in general and Fish & Fisheries in particular.

The '*website*' – '*fishbiopedia.com*', another working platform of the author, will speak more about the achievements and services rendered to the field of education.

www.ingramcontent.com/pod-product-compliance
Lightning Source LLC
Chambersburg PA
CBHW060427220526
45465CB00008B/3043